D0182422

A
Naval History
of the
Civil War

By the Same Author

Third Parties in American Politics
The Forgotten Wars
Stormy Petrel: The Life and Times of General
Benjamin F. Butler

A Naval History of the Civil War

Howard P. Nash, Jr.

South Brunswick and New York:
A. S. Barnes and Company
London: Thomas Yoseloff Ltd

A. S. Barnes and Co., Inc.
Cranbury, New Jersey 08512

Thomas Yoseloff Ltd
108 New Bond Street
London W1Y OQX, England

Library of Congress Cataloging in Publication Data

Nash, Howard Pervear, 1900–
A naval history of the Civil War.

Bibliography: p.
1. U. S.—History—Civil War—Naval operations.
I. Title.
E591.N29 973.75 76-146768
ISBN 0-498-07841-8

Printed in the United States of America

For
Lucinda and Ardith

Contents

Author's Note

If this work seems like a small book about a big subject it is so for two reasons.

One is that the author is a newspaper editor who spends most of his time at his regular work pruning copy so that it will tell a given story in the fewest possible words. This being so, he tends to "write tight," as editors constantly admonish reporters to do.

The other reason is that the author, seeking broadly to describe the parts played by the Union and Confederate Navies in the Civil War, has deliberately dealt only with significant events, not with every pinprick raid, every destruction of a salt works somewhere in the South, every blockade runner captured or driven ashore, etc., or with the orders and reports concerning such things.

A Naval History of the Civil War

1

The Anaconda Plan

In any war the belligerents necessarily adopt strategies shaped by their political aims and military capacities. In the American Civil War the political aim of the Confederate States was to establish an independent nation; the political aim of the northern states was to prevent the South from leaving the Union.

The South, with a comparatively small population and an economy based on agriculture, had no choice except to act defensively. Its only chance to achieve its political aim was to prolong resistance in the twofold hope that the North would finally decide it was not worth the cost to compel the South to stay in the Union and that Great Britain, France, or both, would come to the aid of the Confederacy.

To accomplish its political aim the North had to act aggressively, to attack the Confederate States physically and economically in every possible way and at every possible point.

Militarily the Confederate States were vulnerable to joint Army-Navy attacks because their Atlantic and Gulf coasts were honeycombed with rivers, creeks, inlets, and bays while the Tennessee and Cumberland Rivers extended deep into the country and the Mississippi River bisected it.

Economically the South was vulnerable; having almost no manufacturing capacity of its own, it exported cotton, rice, sugar, and naval stores to New England and Europe in exchange for things as big as steam engines and as small as pins and needles. If this commerce could be stopped the seceding states could not obtain the arms, equipment, and munitions with which to wage a long war.

In view of these facts the political and military leaders of the North quickly agreed upon the importance of interdicting as far as possible the Confederate States' foreign trade. However, there was some discussion among the members of President Abraham Lincoln's Cabinet as to whether this could be done better by declaring that the cities where there had been United States customhouses were no longer ports of entry or by blockading the Confederate States' coasts.

Some of the Cabinet favored the first course because it appeared to be so easy and simple. All it seemed to need was the issuance of an executive order. A fault of this plan was that if a vessel defied such an order and landed its cargo at a closed port it would merely have violated a United States revenue law. Such an offense could be punished only by proceedings in a federal court in the state and district where it occurred, and these courts were in the hands of the Confederates. There were also good reasons to fear that other nations might consider the closing of the ports by fiat to be an attempt to establish a paper blockade. They would certainly protest against this sort of thing and would probably disregard such an order.

Those who favored a blockade pointed out the advantages of this policy: it would close the ports by physical force; thereby effectively putting a stop to the Confederate States' foreign commerce. Furthermore, any vessels violating such a blockade could be captured and condemned in United States admiralty courts or in those of other countries, with the consent of their governments.

Three more or less cogent arguments were advanced against a blockade. To proclaim a blockade is an act of war, hence to blockade the Confederate States would, tacitly at least, grant them the status of a belligerent country. This, in turn, would almost admit the validity of their claim to be an independent nation and might encourage other countries to extend diplomatic recognition to them. (These were things the secessionists hoped would occur; they regarded themselves as patriots, ready, as their ancestors had been in 1776, to wage war against a government under which they were no longer willing to live. But they were things the federal government did not want to have happen; it regarded, and wanted the rest of the world to regard, the Confederates as rebels.) The courts might declare a blockade of some states by other states unlawful, in which event it could not be enforced. Finally, it was said that the United States did not have a navy capable of blockading a coast nearly 3600 miles long.

The last argument was a highly important one. A legal blockade as defined by international law must be an effective one. That is to say it must be maintained by a force capable of actually making access to an enemy's ports difficult and dangerous. When the Civil War began the United States did not possess such a naval force.

Of the 90 vessels of all classes listed in the *Navy Register for 1861,* 21 were unfit to go to sea at all, 27 were laid up in various navy yards in need of more or less extensive repairs or were on the stocks not yet ready to be launched, and 28 were on foreign stations (seven of them as far away as the coast of China). Only the vessels attached to the Home Squadron—the screw steamers *Brooklyn, St. Louis, Wyandotte, Mohawk, Pocahontas, Pawnee,* and *Crusader;* the sailing frigates *Sabine, Cumberland,* and *Macedonian;* the side-wheeler *Powhatan;*

the store ships *Release* and *Supply;* the steam tender *Anacostia*—were in port anywhere in the United States or close enough at hand to be quickly available.

Obviously this handful of vessels could not possibly cover the nearly 200 bays, inlets, river mouths, and deltas along the Confederate States' Atlantic and Gulf coasts. Indeed they could not even effectively blockade the 10 ports—Norfolk, Virginia; Newbern, Beaufort, and Wilmington, North Carolina; Charleston, South Carolina; Savannah, Georgia; Fernandina and Pensacola, Florida; Mobile, Alabama; New Orleans, Louisiana—through which most of the South's imports and exports would have to pass, because they were the only places having deep-water harbors and adequate rail connections with the interior.

Before a decision was made as to what course to follow, Secretary of State William H. Seward discussed the alternatives with Lord Lyons, the British minister to the United States. Lord Lyons made it clear that closing the ports by executive order might lead Great Britain to recognize the Confederate States in order to protect the legitimate rights of British merchants. However, he had to admit that his country could not reasonably object to a blockade since it had often used this form of naval warfare.[1]

In these circumstances President Lincoln proclaimed a blockade of the Confederate States' coasts from North Carolina to the Rio Grande on April 19, 1861, a week after the Civil War began. (When Virginia seceded a short time later the blockade was extended northward to the Potomac River.)

A great many southerners, including Jefferson Davis, President of the Confederate States, were confident that

1. Bern Anderson, *By Sea and By River* (New York: Alfred A. Knopf, 1962), pp. 26–27.

Great Britain would never allow the Union Navy to establish a blockade. This assumption was based on what seemed to be logical grounds. Five million Englishmen relied on the textile industry for their livelihood. England depended almost absolutely on the southern states for cotton; ergo, the British government would not let anything interfere with the free flow of cotton. And if Great Britain, with its incomparable Navy, said there was to be no blockade there would be none.

If the theory, expounded by a North Carolina senator in 1858, that cotton was a king against whom no power on earth would dare to make war[2] ever had any merit, the South chose a poor time to test it. When the Civil War began most of the bumper crop of cotton produced in 1860 had already been shipped. With enough cotton on hand for the country's needs and not foreseeing a long war Great Britain did nothing directly to interfere with the blockade. Thus the Union was able by the end of 1861 to improvise a fleet capable of maintaining an effective blockade and, in cooperation with the Army, to occupy several Confederate ports along the Atlantic coast. To do this the Navy Department chartered or bought and armed more than 100 freighters, passenger liners, coasters, yachts, excursion boats, ferryboats, and tugboats driven by screw propellers, side-wheels, or sails. (All that Gideon Welles, the energetic Connecticut Yankee who headed the Department, asked of these "instant warships" was the ability to float and carry some guns.)

If the United States had found itself at war with a first class naval power in 1861 the lack of ships would have been dangerous, if not catastrophic. In the actual circumstances the Navy's weakness proved to be no great misfortune. The existing fleet consisted chiefly of general-

2. *Congressional Globe,* 1st Sess., 35th Congress, p. 961.

purpose oceangoing vessels; the need of the moment was for craft of other types, some of them highly specialized. Blockade duty called for steamers able to ride at anchor in all sorts of weather, to get underway quickly, and to show a good turn of speed. Some fast cruisers were needed to hunt for Confederate commerce raiders, and, in case of foreign complications, to do some commerce raiding themselves. Harbor fortifications could be attacked with any hope of success only by heavily armed, well-protected ships. Work in the rivers and sounds along the southern coasts demanded highly maneuverable, shallow draught steamers.

As soon as the war began the Navy Department recalled the vessels on foreign stations (it took some of them two years to get back to the United States) and adopted a number of measures to meet the needs of the hour.

The first step was to buy or charter everything afloat that could possibly be used by the Navy. In the beginning purchases were made and charters were arranged directly by the Navy Department or by officers acting under its direction. In July 1861, by which time 12 steamers had been bought and nine others chartered, the Department decided that such transactions, being purely of a mercantile nature, ought to be put in the hands of a businessman. Accordingly Welles appointed his brother-in-law, George D. Morgan of New York, as purchasing agent for the Navy. This choice inevitably led to charges of nepotism against Welles, but they were not well-founded. The vessels Morgan bought were inspected by boards of officers, and the Navy secured—so far as they were available—useful craft at reasonable prices. However, the boards could not set too high standards of excellence or even of fitness because the need for shipping was so desperate.

The craft thus obtained varied in size and sort from

ferryboats and tugs to 2000-ton screw steamers and side-wheelers. Some of the least promising of them did good service. One of the most valuable prizes of the war, the blockade runner *Circassian,* was taken outside of Havana by a former ferryboat.

Another of the Navy Department's emergency measures was to authorize the construction of 16 sloops of war, half of them 1000 tonners, half of them larger. These were long, narrow wooden steamers with fine lines and shallow draught for their size in which other qualities were largely sacrificed for speed. Thus they were suitable for hunting Confederate raiders or themselves becoming commerce raiders. Among them was the *Kearsarge* which sank the C.S.S. *Alabama* a few miles outside of Cherbourg, France, in one of the most famous engagements of the Civil War.

Without waiting for Congress to convene in special session on July 4, 1861, the Navy Department entered into contracts with a number of privately owned shipyards for the construction of 23 screw-propelled, 500-ton gunboats carrying from four to seven guns each. Because the first of these craft were built in less than four months they came to be known as "90 day gunboats."

To procure vessels specially suitable for service in narrow waters the Navy Department had a number of "double-enders" built. These unique warships displaced 300 or 400 tons each and, like ferryboats, their side-wheels were located exactly amidships and they had rudders and pilothouses at both of their ends. Thus they could change course by as much as 180 degrees without having to turn around. However, they proved to be difficult to steer or maneuver, especially in a tideway.

When Welles reported to the Congress that he thought the emergency had made it his duty to take the steps he had to expand the Navy the legislators approved his acts and authorized him to buy more vessels. Before the war ended the Navy Department bought 313 steamers and 105 sailing craft.

While the lack of ships was still an acute problem the Union Navy Department acted on an idea conceived by the Confederates that the easiest way to blockade a harbor would be to sink some vessels in its channel.

On January 11, 1861, three months and a day before the Civil War began, the governor of South Carolina caused four hulks to be sunk in the Main Ship Channel at Charleston in hope of forestalling any further attempts to land supplies or reinforcements at Fort Sumter such as had been made by the S.S. *Star of the West* a short time earlier.

Since no Union vessels attempted to pass these obstructions their effectiveness was never really tested. However, Major Robert Anderson, the commanding officer at Fort Sumter, thought they were formidable and Captain John G. Foster of the United States Army Corps of Engineers said that pilots who knew exactly where the hulks were located were finding it highly difficult to make their way through the channel.

Reports of this sort apparently led the Union Navy Department to adopt the same method of blockading part of the southern coast. The decision to do this may well have been made by the Assistant Secretary of the Navy Gustavus V. Fox rather than by Welles. Welles, a landsman, relied in matters of strategy and tactics to a large extent upon the advice of Fox, a former naval and merchant marine officer who had left the sea to become a textile mill executive. However, orders for the conduct of particular operations were signed by Welles.

During the summer of 1861, not long after Fox had been promoted from chief clerk of the Navy Department to Assistant Secretary of the Navy, Welles issued orders for a number of vessels loaded with stone to be sunk in the inlets among the islands off the North Carolina coasts between Pamlico Sound and the Atlantic Ocean.

Many naval officers, shipmasters, and pilots familiar with the area doubted the merit of this idea because they knew the bottoms of the inlets were of such light, easily shifted sand that new openings would quickly be formed in place of any that were closed. Nevertheless, Welles insisted on having his orders carried out because he believed the plan would materially aid the operations of the blockading squadron.

Without waiting for this scheme to be tested by time the Navy Department decided to extend it to Savannah. This may have been done at the suggestion of Commander Charles H. Davis of the Union Navy and Alexander D. Bache, Superintendent of the U.S. Coast Survey (forerunner of the U.S. Coast and Geodetic Survey); they are known to have mentioned it in a report sent to the Navy Department. Two Boston merchants also urged the same idea upon Welles in unsolicited letters, both dated October 9, 1861. One of these men said, "the ingress or egress of vessels . . . from either Charleston or Savannah" could be prevented "by sinking in the channels vessels thoroughly and completely loaded with stone." The other offered to "prove to the satisfaction of anyone that hulks loaded with granite blocks, heavy enough to prevent their rising on the swell, will remain stationary for 20 years." He confidently added; "I am prepared to form a company here which will give bonds if required, to close the port of Savannah so that not a wherry [a light rowboat] can get in or out."[3]

How much influence, if any, these letters had on Welles and Fox is uncertain. However, on October 17, 1861, Welles instructed Morgan secretly to procure 25

3. Howard P. Nash, Jr., "The Ignominious Stone Fleet," *Civil War Times,* June 1964, p. 46.

old vessels, to arrange for their delivery off Savannah, to have them fitted with pipes and valves designed quickly to flood their holds, and to have them as heavily loaded with blocks of granite as "their safe transmission down the coast" would permit.[4]

The "Stone Fleet," as it quickly came to be called, consisted mostly of old whaleships (there were also a few merchantmen) from New Bedford, Massachusetts, New London, Connecticut, Sag Harbor, New York, and New York City. Local agents shipped their crews—25 captains, 50 mates, 25 stewards, and 239 seamen—and fitted them out in accordance with the Navy Department's specifications.

To prepare them to be sunk easily and quickly large holes were drilled through their sides at places that would be several inches below their loaded water lines. These openings were closed with loose fitting plugs, held in place by bolts with their heads outside, the nuts inside of the hulls. In case these improvised seacocks failed to work properly two large augers were put on board every vessel.

Meanwhile, teamsters gathered up about 7500 tons of stone for which the government paid 50 cents a ton, a price that made it profitable for farmers and householders in the staging areas to tear down many of their stone walls. Because time was important rough stone was taken instead of dressed blocks and the New Bedford vessels were partly loaded with cobblestones intended for a street paving project.

On November 20, 15 vessels of the stone fleet departed from New Bedford, eight from New London, one from Sag Harbor, and one from New York. Although they were buffeted by a gale off Cape Hatteras, they all arrived at Savannah early in December.

---

4. Ibid., pp. 46–47.

The appearance of so many ships led the Confederates to abandon the defenses of Wassaw Sound and Tybee Island, thus permitting such a strict blockade of Savannah to be established as to make it pointless to sink any blockships there.

Three of the stone fleet's vessels were, therefore, used to form a breakwater and a wharf to facilitate landing troops at Tybee Island, four were turned over to the Quartermaster Corps, one became a naval storeship, two sank accidentally, and Davis was ordered to sink the remaining 15 in the Main Ship Channel at Charleston.

Because Davis had been a member of a commission which had surveyed Charleston harbor in 1852 with a view to widening and deepening its channel he was particularly well fitted to carry out the task at hand.

In the light of his knowledge of the harbor he decided to place the vessels to be sunk in three rows, like checkers at the beginning of a game, instead of using them to form a solid wall. Such an arrangement he thought would create a series of shoals around which the tidal currents would swirl and eddy, thus creating a labyrinth through which it would be impossible to navigate.

While he waited for the stone fleet to reach Charleston, Davis had soundings taken to locate the channel from which all buoys had been removed by the Confederates whose pilots could find their way in and out of the harbor without their aid. Because this activity was followed by the arrival, on December 18, of most of the stone fleet plus five gunboats and three transports, the people of Charleston seem to have supposed the city was about to be attacked. During the night the lighthouse was blown up by the Confederates.

Davis's covering force was too powerful to be opposed by the few armed tugs available at Charleston and the channel was out of range of the guns in the various Confederate forts so the sinking of the stone fleet was not at all dangerous. However, the work could be done only at high tide and it took a couple of days to finish it.

Late in January 1862, 20 more vessels were sunk at Charleston in Maffit's Channel. At first this passage had not seemed worth blocking because it was comparatively shallow (12 or 15 feet deep), but it proved to be too useful to blockade runners to be disregarded.

The hopes of many northerners and the fears of many southerners that the stone fleet would put an end to Charleston's days as a major seaport were not fulfilled. The hulks became infested with teredoes (shipworms) which caused them to break up quickly, and tidal currents flowing around the piles of stone soon scoured out deeper channels than those that had been closed.

Welles and Fox must have been upset by this fiasco, but they never admitted it. In fact, Welles did not even mention the stone fleet in his departmental report for 1862.

If the stone fleet had achieved the hoped-for results, many men in and out of the Navy would have claimed credit for having suggested it. As things turned out almost everybody who had anything to do with it followed Welles's example of not discussing it. Thus the story has been almost wholly forgotten except by a few local historians in New Bedford, New London, Sag Harbor, Savannah, and Charleston.

Early in the war the Union high command turned its attention to the Mississippi River which, with its tributaries, formed a sort of inland sea upon whose waters much of the South's interstate commerce moved. (With few roads and fewer railroads south of St. Louis river steamboats were the principal carriers of freight and passengers in the area.)

In May 1861, before even a token blockade had been established anywhere except at Hampton Roads, Virginia (near Norfolk), General Winfield Scott, general in chief of the United States Army, suggested the establishment of a chain of fortified positions extending down the Ohio and Mississippi Rivers from Louisville through Paducah, Kentucky, Memphis, Tennessee, Vicksburg, Mississippi, and New Orleans to the Gulf of Mexico. Scott thought the war would last for three years instead of ending within the three-month period expected by almost everybody else in the North. And, he said, such a cordon as he proposed, "in connection with the strict blockade of the coast" (then being planned) would "envelop the insurgent States and bring them to terms with less bloodshed than by any other plan."[5]

Northern fire-eaters, shouting "On to Richmond," derided Scott's "Anaconda Plan" for strangling the Confederacy, but it was the strategy that won the war. And if it had been followed more closely, with more reliance on the Navy to win by attrition, less on the Army to capture territory as if it were playing a game of football instead of fighting an enemy, the Union would have won the war just as surely, but at far less cost in lives and money than actually was the case.

Because fortifications would have to be overcome at most of the places mentioned by Scott, gunboats would be needed to fight them. In these circumstances the War Department decided to organize a Western Flotilla of such craft.

Naval architecture was, of course, not a subject with which soldiers were familiar, so the Army called upon the Navy Department for help in forming the Western Flotilla. The Navy responded by sending Commander

---

5. *Official Records of the Union and Confederate Armies* (Washington, D. C.: Government Printing Office, 1890–1901), Ser. 1, 51: 369; *Civil War Naval Chronology* (Naval History Division, Office of the Chief of Naval Operations, Navy Department, Washington, D. C.: n. d., pt. 1, p. 12. (Hereafter cited as *NC*.)

John Rodgers to Cincinnati to assist the commanding officer of the Department of the West in buying and building the right kind of boats. (Incidentally, western river craft were, and are, always called boats, never ships, no matter how big or small they may be.)

The few river steamers Rodgers found for sale were poorly suited for military use because, to keep their draught shallow, they had their boilers on deck, where they and all of their steam connections were exposed to damage by gunfire and their crews were quartered in high, thin-walled deckhouses.

Boats of any kind were urgently needed, so Rodgers bought the side-wheelers *Conestoga, Lexington,* and *A. O. Tyler* and converted them into gunboats of sorts by having their deckhouses covered with oak planking five inches thick, their boilers dropped into their holds, their steam pipes lowered as much as possible, and some cannon mounted in them. The *Conestoga* carried two 32-pound smoothbores in broadside with a lighter piece at the stern; the *Lexington* had four 64-pounders in broadside and two 32-pounders; the *Tyler,* six 64-pounders in broadside and a 32-pounder astern.

The War Department also contracted with James B. Eads, a famous civil engineer, for the construction of seven gunboats all of the same model. (Eads, who was an advocate of the Anaconda Plan, may have suggested the formation of the Western Flotilla.)

These boats—the *Cairo, Carondolet, Cincinnati, Mound City, Louisville, St. Louis,* and *Pittsburg* (as the name of the Pennsylvania city was then spelled), named for cities along the rivers where they were to fight—were launched in October and November 1861. They were almost 200 feet long, more than 50 feet in beam, and drew 6 feet of water. Their guns, magazines, living quarters, etc., were enclosed in casemates (with their sides sloping inward at an angle of 45 degrees to deflect projectiles hitting them), armored with 2½ inches of iron backed by 24 inches of oak at their forward ends, but with no backing along their sides or across their sterns because they were expected generally to fight bows on. Since side-wheels could too easily be damaged and stern-wheels would make it difficult to mount stern guns their paddle-wheels were located inside of their casemates, about 60 feet forward of their fantails. (In this regard they were throwbacks; the earliest steamboats on the western rivers were center-wheelers, not stern-wheelers.) They were armed with six or seven 32-pounders, two or three 8-inch Dahlgrens (smoothbore guns named for their inventor, Commander—later Admiral—John A. Dahlgren), and four 42-pound Army rifles.

Eads also converted two river steamers into gunboats. One of them, the *Benton,* was originally a catamaran 202 feet long, 75 feet in beam, drawing 9 feet, built for straddling snags and pulling them from river beds. The space under her hulls was planked over to form a solid bottom and the space above them was made into a single deck. Her casemate had 30 inches of oak across its forward end and 12 inches of oak along the sides and across the stern, covered with 3 inches of iron forward, 2½ inches on each side of the engines, and ⅝ of an inch elsewhere. With a battery of seven 32-pounders, two 9-inch Dahlgrens, and seven 42-pound rifles, the *Benton* was the most heavily armed river gunboat used by either side during the war. Carrying so much armor and so many guns, she was slow and unwieldy. Her top speed in slack water was less than 6 miles an hour; she could not back upstream against the current of the Mississippi River, even when carrying steam at a pressure of 145 pounds to the square inch, and it took her almost 10 minutes to turn around.

Eads's other conversion, the *New Era,* renamed the *Essex,* a steamboat of the same dimensions and armor as the *Benton,* was armed with a 32-pounder, one 10-inch, and three 9-inch Dahlgrens.

Because Naval Constructor Samuel M. Pook helped to design these craft and their casemates somewhat resembled carapaces they were often called "Pook turtles."

These gunboats and others built later for the Western Flotilla played a key part in General Ulysses S. Grant's series of campaigns which, beginning in February 1862, ultimately gained control of the Mississippi River and decisively influenced the outcome of the Civil War.

Early in 1862 the War Department contracted with Charles Ellet, Jr.—another well-known civil engineer—for the conversion of four side-wheel and three stern-wheel high-speed (18 to 20 miles an hour with the current) river towboats into rams.

Ramming was much used by Greek and Roman galleys, but when sails replaced oars this tactic was almost forgotten. Ellet suggested reviving this device after the 250-ton S.S. *Vesta* accidentally rammed and sank the 2794-ton S.S. *Arctic* in September 1854. Unable to interest the Navy Department in this idea he attempted to call it to the attention of the public by publishing (in December 1855) a pamplet entitled *Coast and Harbor Defenses, or the Substitution of Steam Battering Rams for Ships of War,* in which he argued that steamers could be strengthened enough to make them capable of sinking war vessels blockading a harbor.[6]

Ellet harped on this subject for seven years without effect on either the Navy Department or the northern public. However, some southerners in important positions were impressed by his logic and a number of Mississippi River boats were equipped as rams soon after the Civil

6. John S. C. Abbott, "Charles Ellet and His Naval Steam Rams," *Harper's New Monthly Magazine* 32 (February 1866): 297 ff.

War started. On learning of the existence of these craft the Union War Department decided it had better have some rams too.

The boats Ellet chose were strengthened with timber bulkheads 12 to 16 inches wide, 4 to 7 feet high, running from bow to stern; they had iron tie rods so arranged as to prevent their hulls from spreading and iron stays to keep their engines and boilers from toppling when they struck another vessel.

The Western Flotilla's boats were manned by soldiers and officered by the Navy. The friction inevitably caused by this mixture was made worse because the Army insisted that its officers could give binding orders to naval officers of corresponding rank—e.g., an Army captain could give order to a Navy lieutenant, a major to a commander. To add to the confusion the Army Ram Fleet—commanded by Ellet, who was commissioned a colonel—was independent of the Western Flotilla.

After the Confederacy was formed, but before the Civil War began, 10 first-class ships owned by the British East India Company, suitable for conversion into men-of-war, were offered to the Confederate government. These vessels could have been bought, armed, fitted out, and delivered on the southern coast for about $10 million, the equivalent of 40,000 bales of cotton. This proposal was submitted to the government, but its importance was not appreciated and it was rejected by the executive.[7]

This opportunity to procure a ready-made fleet having been missed, the Confederate States began the Civil War with a Navy consisting of six revenue cutters, a steam tender, a few coastwise steamers, and two Coast Survey

7. Robert U. Johnson and Clarence Buel (editors), *Battles and Leaders of the Civil War* (New York: The Century Co., 1884–87), 1: 106–7.

steamers seized by the various states when they seceded. To service this small fleet the Confederates had only two navy yards, almost no privately owned shipyards, one foundry capable of casting big guns (the Tredegar Iron Works in Richmond), two rolling mills fit for heavy work (both of them in Tennessee), and a handful of capable officers who had resigned from the U. S. Navy. In addition they had practically no merchant marine from which to recruit more officers and seamen. Nonetheless, the Confederate Secretary of the Navy, Stephen R. Mallory of Florida, accomplished much with the limited resources available to his country.

A former chairman of the United States Senate Committee on Naval Affairs and a lifelong lover of boats, Mallory was probably better fitted for the position he occupied than any other member of the Confederate Cabinet with the possible exception of the Secretary of the Treasury, Christopher C. Memminger.

Unlike most southerners, Mallory was aware that King Cotton might not be a powerful enough monarch to force Great Britain to aid the Confederacy and he realized how badly a blockade could hurt the country. To break the blockade and cope with the Union Navy he said the Confederate States needed

> fifty light-draft [*sic*] and powerful steam propellers, plated with 5-inch hard iron, armed and equipped for service in [southern] waters, four iron or steel-clad single deck, ten gun frigates of about 2,000 tons [each], and ten clipper propellers with superior marine engines, both classes of ships designed for deep-sea cruising, . . . 3,000 instructed seamen, and 4,000 ordinary seamen and landsmen, and 2,000 first rate mechanics.[8]

He must have realized from the first that this was an impossible dream. But he hoped to procure some armored vessels because he believed that by "fighting with

8. *NC,* pt. 2, p. 28.

iron against wood" the Confederate Navy could offset "inequality in numbers."[9]

When Mallory tried to obtain some ironclads from Great Britain or France he found that the political situation in Europe was such that neither of those countries would sell any of such craft. Thus the Confederate States were thrown back on their own meager resources. Despite extreme difficulties they built several ironclads, some of which almost succeeded in breaking the blockade at a few ports.

With Mallory's encouragement the Confederate Navy Department developed a number of ingenious devices to offset its lack of standard equipment. In this connection the Department established a Torpedo Bureau and a Naval Submarine Battery Service. The torpedoes (they would be called mines today since they were not self-propelled) designed by the Bureau and the Service were, of course, crude things by modern standards. Some of them were, for example, beer kegs, demijohns, tin or copper canisters filled with gunpowder to be detonated either by contact or by means of an electric current sent from batteries on shore. Some were mounted on pilings in harbors and rivers. Such devices sank or disabled 32 Union vessels, and it can be taken for granted that they had a deterrent effect on the captains of other ships, although its extent cannot be estimated.

Colonel Ellet proposed a method of clearing torpedoes from harbors or rivers by attaching

> to the bow of a swift and powerful steamboat a strong framework, consisting of two heavy spars, 65 feet in length, firmly secured by transverse and diagonal braces and extending 50 feet forward of the steamer's bow. A crosspiece [*sic*], 35 feet in length, was . . . bolted to the forward extremities of these

9. *Official Records of the Union and Confederate Navies in the War of the Rebellion* (Washington, D. C.: Government Printing Office, 1894–1922), Ser. 2, 2: 67.

> spars. Through each end of this crosspiece and through the center . . . heavy iron rod[s], 1½ inches in diameter and 10 feet long, descended, . . . terminating in . . . hook[s]. . . . Intermediate hook[s] were attached to each bar 3 feet from the bottom. The three bars were strengthened by a light piece of timber halfway down, through which they were passed and bolted.[10]

This device, Ellet believed, would dispose of torpedoes either by tearing them loose from their moorings or exploding them harmlessly. Individuals and the Navy Department also designed and built several torpedo boats and one true submarine.

The first torpedo boat was named *David* and the name was conferred on the class. These craft were cigar-shaped steamers about 50 feet long. Except that they were smaller they looked much the way a World War I submarine would if it had a smokestack where a conning tower ought to be. Their torpedoes were explosive containers mounted at the ends of booms projecting from the boats' bows. In making an attack the boom was lowered, the torpedo placed against the hull of an enemy ship below the water line and detonated by an electric current or a trigger and a percussion cap. (The expression "to lower the boom" may be derived from this *modus operandi.*)

Mallory was also responsible for an aggressive commerce-raiding campaign conducted by lightly armed but speedy vessels. He hoped these cruisers would raise enough havoc with northern shipping to compel the Union Navy to detach a good many ships to pursue them. Because Welles wisely refused to weaken the blockade the Confederate cruisers had no real effect on the outcome of the war.

In an effort to supplement its limited naval resources the Confederate government invited applications for pri-

10. *NC*, pt. 3, p. 4.

vateering commissions, as the United States had done in the Revolutionary War and the War of 1812. Mallory and some others thought that many persons in Europe as well as the South would gladly avail themselves of the opportunity to profit by preying on United States shipping, but this did not prove to be the case. According to extant records only 53 vessels were ever commissioned as Confederate privateers—most of them were commissioned in 1861—and a mere handful of them actually took any prizes.[11]

(The Union government briefly considered issuing letters of marque and reprisal, but decided against doing so because, as Welles said in a letter he wrote to Secretary of State Seward on October 1, 1861, it would amount to "a recognition of the assumption of the insurgents that they are a distinct and independent nationality.")[12]

The Confederate privateers enjoyed only a limited degree of success because the growing efficiency of the blockade made it increasingly difficult for them to send their prizes into their own country's ports and the ports of other nations refused to permit their entry. This was true even of British and British colonial ports, although a neutrality proclamation issued by Queen Victoria at the beginning of the war was interpreted in many other ways so as to favor the Confederate States.

Those who engaged in privateering were motivated more by the hope of profit than by patriotism and most of them quickly turned to the far more lucrative business of blockade running.

The privateers had even less effect on the war than the Confederate cruisers did. (The accomplishments, such as they were, of the cruisers and the privateers are described in chapters 15 and 16.)

11. Ibid., pt. 6, pp. 328 ff.
12. Ibid., pt. 1, p. 28.

Blockade running, on which the Confederate States had to rely almost absolutely for such necessities as arms, ammunition, and medicines and for such luxuries as coffee and tea, was ridiculously easy at first. For a while even sailing vessels were able to evade the few Union ships stationed off the Confederate ports. However, as the blockade became stricter, methods of evading it had to be improved. For a time blockade runners actually bound for ports in the Confederate States cleared for fictitious destinations, such as Nassau or Bermuda. The capture and condemnation of several vessels that tried this scheme quickly brought it to an end. It was replaced by a system which used two ships; one legally to cross the Atlantic to Bermuda, Nassau, or Havana, the other to make the short, but illegal, run into a Confederate port.

The vessels used for the latter purpose were typically long and narrow (about nine times as long as they were wide) side-wheelers with feathering paddles and one or two raking funnels capable of being lowered close to the deck (for which reason they had forced draft boilers). Their freeboard was low and they were painted dull gray to minimize visibility. They had only two short masts with crows-nests on them and turtleback forecastle decks to enable them to drive through fairly heavy sea. To avoid telltale smoke they burned anthracite coal if it could be had; Welsh coal was an acceptable substitute, but soft coal was never used.

Usually a blockade runner's departure for a Confederate port was timed to bring her there on a moonless night at or about high tide. A sharp lookout was kept for ships on the outside station. When a blockade runner was discovered she relied on speed and evasive action to escape, disregarding gunfire unless actually hit and disabled. Once past the outer line of vessels and near her destination a blockade runner either hid in some inlet to wait for darkness or dashed into port according to circumstances.

In this game the blockaders were at a distinct disadvantage; they could choose nothing, but had to wait on events in a state of constant watchfulness combined with prolonged inaction, about as trying conditions as it is possible to imagine.

Throughout the war blockade running paid owners and crews extremely well. At first owners were satisfied with profits of 100 to 200 percent; later they made from 1500 to 2000 percent in depreciated Confederate money. As for the crews, rates paid for a trip from Nassau to Wilmington and back were as follows (with half paid before leaving port, the balance on returning there:

Captain	$5000
Chief Officer (1st Mate)	1250
Second and Third Officers	750
Chief Engineer	2500
Seamen and Firemen	250
Pilot	3750[13]

Unfortunately for the Confederacy blockade running was left almost entirely in private hands and was virtually unregulated by the government. The few vessels owned by the government brought in large (but not large enough) amounts of arms, ammunition, and supplies. But most blockade runners were owned by men who were more interested in profits than in the Confederate cause and they all too often carried goods that were expected to sell at high prices (Madeira wine, French brandy, crinoline dress material, etc.) rather than matériel of war. A large number of caricatures of Major General

13. James Russell Soley, *The Blockade and the Cruisers* (New York: Charles Scribner's Sons, 1883), p. 167, where the figures are expressed in British pounds which have been converted at the then-average rate of $5 to the pound.

Benjamin F. Butler, who was hated throughout the South, were found in one captured blockade runner. If they had been safely delivered they undoubtedly would have amused some southerners, but nobody can have imagined they would have contributed anything toward winning the war.

2
First Blows

Before the Civil War was many days old the Confederacy added greatly to its military potential by seizing the Gosport navy yard (often miscalled the Norfolk navy yard) located in Portsmouth, Virginia, across the Elizabeth River from Norfolk, about 12 miles above Fortress Monroe situated at the entrance to Hampton Roads.

This yard, in which the government had approximately $10 million invested, was the most valuable and the most extensive establishment of its kind then existing in North America. Covering an area three-quarters of a mile wide and a mile long, it contained machine shops, carpenter shops, sail lofts, gun lofts, etc., with all of their tools and appurtenances, and a drydock capable of handling the largest ships afloat. Because of its nearness to Washington it had long been the country's busiest navy yard and principal storage depot. When the Civil War began about 1200 cannon of various calibers, a great many small arms, a large amount of gunpowder, and several tons of food were stored there; the line-of-battleships *Pennsylvania, Delaware,* and *Columbus,* the sailing frigates *Columbia, United States,* and *Raritan* were laid up there; the brig *Dolphin,* the sailing sloops *Germantown, Plymouth,* and *Cumberland* were there waiting for orders, and the steam frigate *Merrimack* was there to have her engines overhauled after a voyage from the West Coast around Cape Horn. (This vessel is often misspelled *Merrimac:* the *Navy Register for 1861,* the last in which her name appears, spells it with a final "k.")

The Secretary of the Navy was aware that if Virginia seceded the Gosport navy yard, with all of the vessels and stores in it, was in danger of seizure, as much other federal property had been taken by other states. Therefore he sent 250 men there, transferring them from the Brooklyn navy yard on March 31, 1861. When the war began he appealed to General Scott for enough troops to defend the place. On being told by Scott that none of the few soldiers in Washington could be spared for duty anywhere else Welles wrote confidentially to Captain Charles S. McCauley, the commanding officer of the Gosport yard, advising him, in view of recent events, to exercise great vigilance in guarding the public property in his charge, but to avoid doing anything that would antagonize the authorities of Virginia. Until the last moment President Lincoln hoped this state might not secede.

In a letter, dated April 10, 1861, as well as in other messages sent to McCauley on the same day, he was specifically instructed to prepare the *Merrimack* for an early departure. He was also told that it would be a good idea to order as many as possible of the rest of the vessels at the yard either to put to sea or to leave for other stations, to keep the Navy Department informed of conditions at Norfolk, of any trouble he might apprehend, and to ask for any additional force he might think necessary for the security of the yard.

The following day he was again specifically instructed to get the *Merrimack* ready to go to sea as quickly as possible. His reply to this order, that four weeks would be needed to repair the frigate's engines, was discredited by B. F. Isherwood, Chief of the Bureau of Steam Engineering. Isherwood, therefore, requested and received an order to proceed forthwith to the Gosport navy yard to attend personally to the task of repairing the vessel's engines.

Reaching the yard late Sunday, April 14, Isherwood found the *Merrimack*'s engines almost completely dismantled and learned that nearly all of the civilian employees had quit their jobs. He persuaded a few mechanics to return to work and supervised them so efficiently that he was able to report to McCauley at four o'clock Wednesday afternoon that the *Merrimack*'s boilers were ready to be fired. McCauley, anxious not to alarm the Virginia authorities, told Isherwood to wait until the next morning before he started the fires. Isherwood set regular engine room watches throughout the night and lit the fires at daybreak. By this time McCauley had been directed by the Navy Department to have the *Merrimack*'s guns put back on board the ship, to have the *Germantown, Dolphin,* and *Plymouth* made ready to be towed away from the yard, and to have as much matériel as possible put into these vessels for removal beyond danger of seizure. Nevertheless, he refused to let the *Merrimack* go when Isherwood reported at 9 A.M. that she had steam up and was ready to depart.

McCauley adhered to this decision with the stubbornness sometimes displayed by weak-willed men even after Isherwood called attention to his peremptory orders to remove the *Merrimack* at the earliest possible moment. At 2 P.M. McCauley ordered Isherwood to draw the ship's fires. Isherwood flourished his orders at McCauley once more, but the Captain insisted on keeping the *Merrimack* tied up. Waiting only long enough to have McCauley put his order to detain the *Merrimack* in writing, Isherwood left for Washington where he told

Welles that in his belief McCauley was "under the influence of liquor and bad men."[1]

Up to this time Welles had considered McCauley to be a good and faithful officer, though not a particularly capable one.[2] In the light of Isherwood's report of how things stood at the Gosport navy yard Welles decided that McCauley would have to be relieved. Accordingly Captain Hiram Paulding was ordered to proceed forthwith from Washington to Norfolk in the 15-gun U.S.S. *Pawnee*—a second-class sloop which had just returned from a cruise and discharged her crew—with as many officers and seamen as he could quickly gather. He was instructed, on reaching Norfolk, to take charge of all of the naval vessels there, to repel force with force if necessary, and to the best of his ability to protect all public property there from seizure. "On no account," his orders read, "should you allow the arms and munitions [at the Gosport navy yard] be permitted to fall into the hands of insurrectionists, . . . and should it become necessary, you will, in order to prevent that result, destroy the property."[3]

Leaving Washington late in the afternoon of Friday, April 19 (the day of the so-called Baltimore riot in which four men of a Massachusetts Volunteer Militia regiment were killed on their way to the capital) the *Pawnee* reached Fortress Monroe the following afternoon. Here she stopped long enough to pick up the Third Massachusetts Regiment, which had arrived at Fortress Monroe that morning. Although the soldiers were embarked so hastily that no rations were issued to them, it took so long to get them on board that the *Pawnee* did not reach Norfolk until 8 P.M.

---

1. Gideon Welles, *Diary of Gideon Welles* (Boston: Houghton, Mifflin Company) 1: 44.
2. Ibid., 1: 43.
3. *Official Records of the Union and Confederate Navies in the War of the Rebellion* (Washington, D. C.: Government Printing Office, 1894–1922), Ser. 1, 4: 282. (Hereafter cited as *ORN*.)

A couple of hours before the *Pawnee* arrived at the Gosport navy yard McCauley had ordered all of the vessels there except the *Cumberland* to be scuttled. He did this because he had been led by a *ruse de guerre* to believe that several thousand Virginia militiamen were poised for an immediate attack on the yard. Throughout the day trains actually carrying the same civilians back and forth had been running into and out of Norfolk with the men cheering loudly as they came into town and departing quietly.[4]

McCauley had also seen some tugs moving about near the yard during the night and he thought they were towing vessels into positions to be sunk to block the channel from the navy yard to the Elizabeth River. This assumption was correct, but there was room enough for the *Merrimack* to have passed this barrier.

Paulding planned to defend the yard with the *Pawnee* and the vessels there. On finding only the *Cumberland* in any condition to fight and, perhaps frightened by McCauley's estimate of the number of Virginia troops in the vicinity, Paulding quickly abandoned this idea and decided to complete the destruction of the property begun by McCauley.

Working by the light of a full moon one demolition party, commanded by Captain H. G. Wright and Commander John Rodgers, mined the drydock; another, under Commander James Alden, set fire to the storehouses and shops; a third group, led by Commander Benjamin F. Sands, fired the shiphouses; and Lieutenant Henry Wise and his men undertook to burn the sinking ships. (If they had merely been sunk they could easily have been raised; Paulding sought to prevent this by burning them. However, the men who scuttled them had done their work too well. Several of them, including the *Merrimack,* went down so fast their hulls were scarcely

4. Virgil C. Jones, *The Civil War at Sea* (New York: Holt, Rinehart, Winston, 1961–62) 1: 88–89.

damaged by fire and it was possible to refloat and repair them. The *Merrimack,* as we shall see, became an important addition to the Confederate Navy.) The attempt to destroy the drydock failed completely because the fuse was either cut by the Virginians who swarmed into the yard as soon as it was abandoned or it fizzled out. Most of the cannon had to be left behind because they could not quickly be moved or destroyed. Two thousand barrels of gunpowder, a large number of shells, nearly 12,000 barrels of ship's bread (hardtack), 1000 barrels of pork, 700 barrels of beef, lesser quantities of flour, rice, sugar, and coffee, and several thousand uniforms also fell into the Confederates' hands.

Welles, who later described McCauley's actions as having been fatally erroneous, charitably said his behavior resulted from his misplaced confidence in his junior officers, all of whom turned out to be Confederate sympathizers.[5]

In fairness to McCauley it must be emphasized that he had been told several times to avoid needlessly alarming the authorities of Virginia, but it remains a fact that deliberate treachery on his part could hardly have hurt his country more than his misjudgment did, for the blow dealt the Union Navy at Norfolk in 1861 was second only to the disaster suffered by the United States Navy at Pearl Harbor 80 years later.

And Paulding is almost, if not fully as much, to blame in the matter as McCauley because he could have used the *Cumberland*'s heavy battery to level Norfolk and Portsmouth in defense of the Gosport navy yard. If Paulding had been unwilling to act this draconically the 300 men of the *Cumberland*'s crew, those of the *Pawnee,* the men from New York, and the Massachusetts regi-

5. *Report of the Secretary of the Navy,* July 4, 1861, as quoted in F. Moore, *Rebellion Record* (New York: G. P. Putnam; D. Van Nostrand, 1861–67) 2, Docs., 236.

ment could have held off the Confederates until the demolition of the yard had been completed and the cannon rendered useless if they could not be removed.

As Admiral Daniel Ammen wrote in 1883: "It is not too much to say now that [the Gosport navy yard] should have been held at any cost of life, long enough at least to have destroyed the cannon, workshops, and ships."[6] Had this been done the Confederates could not have waged so long a war as they did nor have fought so vigorously.

Some of the cannon the Confederates captured at the Gosport navy yard were quickly installed at strategic points along the Potomac River to prevent Union shipping from using that convenient route between Washington and the North. The Union Navy Department reacted to this development by hastily organizing a Potomac Flotilla, which was charged with the duty of keeping the river open and stopping traffic across it between Virginia and Maryland where there were a great many Confederate sympathizers.

Soon after this Flotilla was formed it participated in the first invasion of the Confederate States, a movement across the Potomac from Washington to Arlington Heights and Alexandria, Virginia.

Arlington Heights was only three and a half miles from the capitol, two and a half miles from the White House, and one mile from Georgetown. The Confederates had a detachment quartered in Alexandria; reinforcements from south of there could, in a single night, move into the Heights, destroy the Virginia end of the bridges across the Potomac, erect mortar batteries, and bombard Washington. Obviously a Union force would have to occupy Arlington Heights to protect the capital.

6. Daniel Ammen, *The Atlantic Coast* (New York: Charles Scribner's Sons, 1883), p. 5.

Because of the scarcity of troops in Washington, it was easier to recognize the need to move upon the Heights than to do it, but finally, at 2 A.M., May 24, 1861, three columns crossed the Potomac. Three regiments went by way of the aqueduct at Georgetown; four crossed the Long Bridge from Washington; and one, 24-year-old Colonel Elmer Ellsworth's Fire Zouaves, was ferried from its camp below the city to Alexandria where the *Pawnee* stood by to cover the troops' landing.

These movements were almost unopposed. There were no substantial Confederate forces at the ends of the bridges and the detachment at Alexandria evacuated that place after receiving a demand to surrender made without authorization by the captain of the *Pawnee*.[7]

Soon after Ellsworth's regiment landed he went to the Marshall House, the town's principal hotel, to lower a Confederate flag he had been able to see from his camp. After accomplishing his purpose he and the two enlisted men with him started down the narrow, winding stairs from the roof of the building. As Ellsworth reached the third step above the second floor landing a door was flung open and James T. Jackson, the proprietor of the hotel, sprang out and killed the Colonel with a blast from a double barrel shotgun. Before Jackson could fire the second barrel one of the soldiers shot and killed him.[8]

Young Ellsworth, with whom President Lincoln had become friendly, was buried with imposing honors from the East Room of the White House with the President, the Cabinet, and many high government officials among the mourners.

On May 31, 1861 the Potomac Flotilla's side-wheeler

7. John G. Nicolay, *The Outbreak of the War* (New York: Charles Scribner's Sons, 1883), p. 110; *ORN,* Ser. 1, 4: 478–79.

8. Robert U. Johnson and Clarence Buel, eds., *Battles and Leaders of the Civil War* (New York: The Century Co., 1884–87), 1: 179; Nicolay, *Outbreak,* p. 113.

Thomas Freeborn (carrying two guns), the steam tender *Anacostia* (mounting two guns), and the tugboat *Resolute* (armed with one gun) attacked a Confederate battery at Split Rock Bluff, about 30 miles below Washington where Aquia Creek flows into the Potomac. Both sides fired a good many shots during the two hours this engagement lasted, but neither of them did any serious harm to the other.

Another battle was fought at the same place the following day. This time the *Anacostia* was hit nine times, but not seriously damaged, while the *Thomas Freeborn* had one of her paddle-wheels crippled by a shot and the recoil of her own guns nearly shook her to pieces. After five hours' fighting the vessels' crews were so tired that Commander James H. Ward, the Flotilla's flag officer, broke off the engagement.

The Flotilla's third battle was fought and the Union Navy's first casualty was sustained on June 27 at Mathias Point in the Potomac. At this time Commander Ward went ashore with a small party to clear away some trees, where he thought a concealed battery could be conveniently erected. Almost as soon as he landed he was attacked by several hundred Confederates who appeared suddenly from behind a hill. Ordering the landing party to lie off in its boats, Ward went back on board the *Thomas Freeborn* and opened fire on the Confederate column. He was killed as he was sighting a gun on the forward deck.

Ordinarily the captains of even small vessels, much less flag officers, did not act as gun pointers; Ward did so on this occasion because the piece (of field artillery) he was aiming was mounted on a carriage he had designed and he wanted to test it for himself.

Soon after Ward died his second in command recalled the landing party. When the men on shore undertook to return to the Union vessels their commanding officer,

Lieutenant James H. Chaplin of the *Pawnee,* sent them to the boats, waiting himself to be the last to leave. By the time he came to the beach only one man remained with him and the boats had drifted away from the shore. Chaplin was unwilling to bring one of them back under fire and the man with him could not swim so Chaplin took him on his shoulders, musket and all, and swam to the nearest boat with him.

Most of the Flotilla's landing operations, and there were many of them, were more successful than its first one. The first prize taken by the Flotilla, the schooner *Somerset,* captured by the *Resolute,* was followed by the capture of many other craft ranging from rowboats to fair-sized blockade runners.

Late in August 1861; the Navy Department assigned 200 marines to the Flotilla for the purpose of scouring the Maryland countryside (especially in the vicinity of Port Tobacco) for places suspected of being Confederate depots for provisions and arms to be used for an invasion of Maryland, which of course never occurred.

An action of no great moment in itself, but typical of the sort of work done by the Potomac Flotilla, occurred in January 1863. Two Union barges, the *J. C. Davis* and the *Liberty,* transporting supplies, broke loose from their moorings in Cornfield Harbor, Maryland, and drifted into the Coon River in Virginia. On hearing of this event the commanding officer of the U.S.S. *Dan Smith* ordered a cutter "to rescue the crews and recapture or destroy the boats." This order was carried out by a party headed by an acting ensign.[9]

Early in 1864 the Provost Marshal General of Maryland wrote to the commander of the Potomac Flotilla to advise him that the schooner *Ann Hamilton* ought to be carefully examined because her owners' agent was known

---

9. *Civil War Naval Chronology* (Washington, D. C.: Naval History Division, Office of the Chief of Naval Operations, Navy Department, n. d.), pt. 3, p. 13.

to be a strong Confederate sympathizer. As it turned out the *Ann Hamilton* had already been taken off Point Lookout, Maryland, by the U.S. Revenue Cutter *Hercules,* attached to the Potomac Flotilla. A search of the schooner confirmed the Provost Marshal General's suspicions: large quantities of salt and lye and more than $15,000 in Confederate money were found on board and she was sent to Washington for adjudication.

About five weeks later the *Yankee,* under the command of an acting lieutenant, reconnoitered the Rappahannock River to within a mile of Urbanna, Virginia. He reported: "We learned that there is now no [Confederate] force of any importance at or near Urbanna, although the presence of troops [there] a short time ago was confirmed."[10]

A couple of days after this reconnaissance was made General Butler, commanding the Army of the James, asked the *Yankee* "to watch the Rappahannock from 10 miles below Urbanna to its mouth." The ship's captain was ordered to lend the General as much assistance as he could[11]—a typical example of the co-operation of the Flotilla with the Army.

One of the Flotilla's more spectacular engagements occurred on April 21, 1864 near Urbanna. As the 85-ton, 50-foot-long *Eureka* closed some small boats near the shore of the Rappahannock she was brought under a heavy fire by a large group of Confederate soldiers hidden in some woods. Despite the fact that she had been caught utterly by surprise the *Eureka* fought back and forced the Confederates to withdraw. "It was quite a gallant affair," as the Flotilla's commanding officer remarked, "and reflected a great deal of credit upon both the officers and men of the *Eureka.* . . ."[12]

About three months before the war ended Welles told

---

10. Ibid., pt. 4, p. 30.
11. Ibid.
12. Ibid., pt. 4, p. 47.

the Flotilla's commander to be particularly watchful for Confederate agents en route to Richmond, who would cross the Potomac in rubber boats. "These messengers," Welles added, "wear metal buttons, upon the inside of which dispatches are most minutely photographed, not perceptible to the naked eye, but are easily read by the aid of a powerful lens."[13]

On April 22, 1865 Welles sent word to the Flotilla that "[John Wilkes] Booth [President Lincoln's assassin] was near Bryantown, [Maryland], last Saturday [April 15], where Dr. [Samuel A.] Mudd set his ankle, which was broken. . . . The utmost vigilance is necessary in the Potomac and the Pautunxet to prevent his escape. All boats should be searched."[14]

This alert remained in force until April 27, when Welles wrote to the commanding officer of the Flotilla: "Booth was killed . . . yesterday, 3 miles southwest of Port Royal, V[irgini]a."[15]

The Potomac Flotilla was disbanded July 31, 1865. During the first year of its existence it was unable to keep the river open for Union shipping because its vessels were too small to carry guns heavy enough to cope with those the Confederates could mount on shore and the Union Army either could not or would not help the Flotilla against them. However, it did prevent the Confederates from using the river for their purposes. It also contributed significantly to the Union victory by assuring the safety of General Grant's base of supplies (in 1864 and 1865) at Fredericksburg, Virginia.

Almost as soon as the blockade began it became apparent that because of the frequent need to refill the ships' coal bunkers the Union Navy needed to have a

13. Ibid., pt. 5, p. 10.
14. Ibid., pt. 5, p. 90.
15. Ibid., pt. 5, p. 94.

base or bases a substantial distance south of Hampton Roads. Sending ships to Hampton Roads to refuel reduced the efficiency of the blockade both because of the time they had to be away from their stations and because on the way back to their stations, particularly to those far to the south, they used up part of their fresh supplies of coal. The alternative, coaling at sea, was at best a slow, difficult operation and one that was impossible in stormy weather. Major repairs, of course, could not be made at sea in the best kind of weather.

In June 1861 Welles appointed Captain (later Admiral) Samuel F. Du Pont and Commander Davis of the Navy, Major John G. Barnard of the Army Corps of Engineers, and Alexander D. Bache of the United States Coast Survey as members of a board to consider this problem. (This is an early, if not the earliest, example of a joint chiefs of staff in American history. The broad policies the board set forth were essentially followed until their culmination at Appomatox Courthouse nearly four years later.)

The board was instructed to select two ports, one in South Carolina, the other in Florida or Georgia, as places suitable for the establishment of coal depots and make plans for their capture.

In its first report, made early in July, the board recommended the seizure of Fernandina, Florida, near the Florida-Georgia state line. This place was selected because it was healthy, had ample wood and water available, and its harbor was deep enough to permit all except the biggest ships in the Navy to enter it. If it could be taken suddenly the buildings of the unfinished Florida Railroad could be used for repair shops and the storage of coal. The board estimated that 3000 men could take Fernandina and hold it.

Later in the same month the board recommended that the Atlantic Blockading Squadron be divided into two commands with the boundary line between them to be in North Carolina near the South Carolina border. There were two reasons for this suggestion: the nature of the coast north and south of this line differed considerably and the Atlantic Blockading Squadron had become too big for one man to administer efficiently.

Before the Navy Department had a chance to act upon either of these recommendations Captain Silas H. Stringham, commanding officer of the Atlantic Blockading Squadron, called the board's attention to Hatteras Inlet, a passage between the Atlantic Ocean and Pamlico Sound about 13 miles below Cape Hatteras. It had been opened by a storm in 1847.

As Stringham pointed out (after the fact had been brought to his notice by the masters of several merchantmen) Hatteras Inlet was being used by Confederate privateers and government vessels as a base from which to prey upon coastwise shipping.

And, according to a letter sent to Welles by an unidentified correspondent, "unscrupulous New England men" were clearing vessels for the West Indies and arranging for them ostensibly to be captured off Cape Hatteras. After being unloaded at Hatteras Inlet they went on to the West Indies, loaded another cargo, and were captured again on their way home.[16]

Stringham had learned, as he told the board, that Hatteras Inlet was guarded by two well-designed forts built of logs and sand. The smaller of them, Fort Clark, mounted five 32-pound guns bearing upon the ocean approach to the inlet. The other work, Fort Hatteras, was on an elevation in a position which enabled its 25 cannon of various calibers to command the inlet and to protect Fort Clark. Its octagonal walls enclosed nearly

---

16. *ORN,* Ser. 1, 6: 111.

an acre and a half and contained an extensive bomb-proof. (Incidentally, both of these forts were armed with guns taken from the Gosport navy yard.)

The board recommended closing the inlet by sinking a couple of schooners loaded with stone across its channel. The Navy called upon the Army for help in overcoming the forts to permit this to be done. The Army responded by furnishing 860 soldiers, commanded by General Butler, a tugboat, two steam transports, and a number of iron surfboats to be used as landing craft. (Some of these boats were carried on the schooners' decks, two of them were towed by the warships.) Stringham commanded the naval contingent of the expedition, consisting of the steamships *Minnesota, Wabash, Pawnee, Susquehanna, Monticello,* and *Harriet Lane,* and the sailing ship *Cumberland.*

Calm, pleasant weather prevailed from the time the expedition (or task force, as it would be called today) departed from Hampton Roads early in the afternoon of August 26, 1861, until after it passed Cape Hatteras a little more than 24 hours later. However, by the time the troops were aroused at 4 A.M. August 28, a fresh southwest wind was piling a heavy surf on the beach. Two and a half hours later the schooners, with as many soldiers in them as they could carry, were towed as near the shore as seemed safe, anchored there, and the men began landing under cover of the guns of the *Pawnee, Harriet Lane,* and *Monticello*. After about 300 soldiers, a few marines, and two 12-pound howitzers had been put ashore, rising seas made it impossible to land any more troops or even to send supplies to those already on shore. By this time all of the surfboats and all except one of the small boats loaned by the covering battleships had been swamped, because the moment a boat's bow touched bottom its still floating stern would swing around, causing it either to fill with water or to capsize, and there was grave danger that the schooners with the rest of the soldiers in them might drag their anchors and pile up on the beach. Several risky attempts to pass hawsers to the schooners failed before they were finally towed to safety.

At 10 A.M., while the troops were landing three miles north of Fort Clark, the *Minnesota, Wabash, Cumberland,* and *Susquehanna* opened a brisk fire of grapeshot and canister to keep the Confederates from serving their guns.

Because only about 60 percent of a ship's guns (those in broadside and a few others so mounted that they could fire in any direction) could be brought to bear upon a target at one time, one piece on shore was estimated as being worth five afloat. Thus the forts theoretically had some little advantage over the ships, which mounted a total of 134 guns. Stringham sought to offset the forts' supposed superiority by keeping his vessels constantly in motion, passing and repassing the forts on an elliptical course, thus forcing the Confederate gunners to fire at ever varying ranges. (Naval men were used to doing this sort of thing; their gun platforms were never stationary.)

In this engagement, as was to be the case in many others, the Confederates were plagued by a shortage of artillery ammunition. Less than two hours after the battle began Fort Clark ran out of ammunition and the ships' heavy fire made it impossible for a supply to be brought over from Fort Hatteras. In these circumstances Fort Clark's defenders spiked their guns and retired to Fort Hatteras. At 2 P.M. Fort Clark was occupied by Union troops. A couple of hours later Stringham ordered the *Monticello* to steam through Hatteras Inlet because he though Fort Hatteras had also been abandoned; for some time past it had had no flag flying. The ship had scarcely entered the inlet before she ran aground. She immediately became the target of a galling fire from the fort, which, silent but not silenced, had been waiting for just such an opportunity as it now had. The other vessels soon freed the *Monticello,* then all of them bombarded the fort until 6 P.M. At that time threatening weather caused most of them to haul offshore for the night, leaving three of them to afford what protection they could to the men on shore.

With the weather pleasant and the sea calm the ships closed Fort Hatteras again early the following morning. The *Susquehanna, Wabash,* and *Minnesota* opened fire at 8 A.M.; the *Cumberland* joined them a little later and was taken in tow by the *Wabash.*

While the warships bombarded the forts the tugboat *Fanny,* with General Butler on board, stood in close with the shore to permit him to direct the landing of the rest of the troops. Soon after 11 A.M., before any additional soldiers had landed, the fort hoisted a white flag. Butler signalled to one of his subordinates on shore to find out what the flag meant. This officer brought the General a note from Captain Samuel Barron of the Confederate Navy, who was in command of the defenses of Hatteras Inlet, offering to surrender his force if the officers might keep their side arms. Barron also sent a verbal communication to the effect that there were more than 600 men at the fort with another 1000 less than an hour's march away, but he wanted to avoid further bloodshed. Butler replied, without waiting to consult Stringham: "The terms offered are these: Full capitulation; the officers and men to be treated as prisoners of war. No other terms admissible."[17]

Barron, who had probably been trying a poker player's bluff with his story of many more men near at hand, promptly agreed to surrender unconditionally. Butler

17. Ibid., Ser. 1, 6: 120.

then told him that since the expedition was a joint one the formal capitulation would have to be made to both Union commanders on board the *Minnesota*.

Because Hatteras Inlet offered a safe anchorage in almost any sort of weather and once across its bar shallow draught vessels could reach many places in Virginia and North Carolina, Stringham and Butler thought it was potentially too valuable a place merely to be corked up. Accordingly they disregarded their orders to sink the schooners they had with them there. Butler, who was supposed to have returned immediately to Fortress Monroe with all of his men, left a substantial force at the Inlet; Stringham left the *Monticello* and *Pawnee* there, and Butler made a quick trip to Washington to explain his idea and Stringham's to their superiors. On the basis of Butler's representations the War Department decided to garrison Hatteras Inlet.

This was both a fortuitous and a fortunate decision. At the time it was made the stone fleet had not been sunk at Charleston so there was no way to foresee that if Hatteras Inlet had merely been blocked it would soon have been reopened and the Confederates, of course, could have rebuilt their forts there. As it turned out the place became a highly useful Union base where coal, ammunition and other supplies were stored and from where another strategically important expedition (to be described in the next chapter) was launched.

Although the Hatteras Island expedition is not even mentioned in some histories of the Civil War, and is dismissed with a paragraph or two in many others, it was in fact a highly significant event. It demonstrated the vulnerability of fortifications, the only means of coast defense available to the Confederacy because of its lack of a navy. If Hatteras Inlet could be taken so could other places on the Confederate coasts.

As soon as Hatteras Inlet was fully secured the naval commandant there turned his attention to Fort Ocracoke, on Beacon Island, commanding Ocracoke Inlet, about 20 miles to the southwest. On September 16 he sent the tug *Fanny* with 61 men of the Naval Brigade on board with a launch carrying 22 sailors, six marines and a 12-pound howitzer to the island. This party found the fort deserted with four 8-inch shell guns and 14 heavy 32-pounders left in it. While some men were breaking the trunnions off of these pieces the launch went to the nearby town of Portsmouth where three 8-inch guns were found on the beach and one mounted on its carriage in what was to have been a sand battery. After damaging these guns beyond repair the launch went back to the *Fanny* and the expedition returned to Hatteras Inlet.

Soon after this minor triumph occurred a body of Confederates stationed on Roanoke Island north of Hatteras Inlet captured the *Fanny* in Pamlico Sound and attacked an Indiana regiment occupying Chicamicomico near the northerly end of Hatteras Island. The landing of the Confederates above and below the Indiana troops was delayed by shoal water long enough for the latter to beat a hasty retreat to the vicinity of Hatteras Inlet. Here a regiment sent out from the base at the inlet and gunfire from the frigate *Susquehanna* turned the tables and caused the Confederates to backtrack. The *Monticello* followed them, firing at them across the low sand fields and perhaps annoying them a bit, but not doing them any real harm.

3

Atlantic and Gulf Coasts—1861-62

Soon after Stringham returned from Hatteras Inlet he asked to be relieved as flag officer of the Atlantic Blockading Squadron because he found the duty too strenuous for a man of his age—he had been in the Navy for more than 50 years. As recommended by the blockade board the Atlantic Squadron was now divided into two commands. One of them, designated the North Atlantic Blockading Squadron, headed by Captain Louis M. Goldsborough, was to cover the coasts of Virginia and North Carolina. The other, known as the South Atlantic Blockading Squadron, commanded by Captain Du Pont, was to operate on the coasts of South Carolina, Georgia, and eastern Florida.

When Du Pont took command of the South Atlantic Squadron the blockade board, of which he had been a member, ordered him to choose two of four places—Bull's Bay, St. Helena Sound, or Port Royal, all in South Carolina, and Fernandina, Florida—and to occupy two of them with the aid of the Army.

Du Pont wanted to attack Bull's Bay because he

thought it would be the easiest place to take, but Fox, who had recently been appointed Assistant Secretary of the Navy, urged him to move against Port Royal instead.

Port Royal was by far the better choice. Although the place was little known to most northerners, it was the finest natural harbor on the Confederate States' Atlantic coast. Situated about halfway between Savannah and Charleston, it had a channel a mile wide leading into a bay at the confluence of the Broad and Beaufort Rivers which offered a safe anchorage in even the worst sort of weather and was big enough to contain the entire United States Navy of that day.

Plans were made for a joint Army-Navy expedition to depart from Hampton Roads early in September 1861 with a naval contingent of 25 warships, ranging in size from the *Wabash,* a 3200-ton, 40-gun frigate, down to small steamers carrying only one or two guns apiece, and a number of coal schooners. For its part the Army contributed 50 transports, the U. S. Third Artillery, and about 26,000 troops from Maine, New Hampshire, Rhode Island, Connecticut, New York, Pennsylvania, and Michigan.

The Port Royal expedition's schedule was upset by the battle of Bull Run (or Manassas), fought on July 21, 1861. The Union setback on that occasion caused so much fear to be felt for the safety of Washington that the troops intended for the expedition were detained to defend the capital.

Finally, on October 28, the coal schooners, convoyed by the sloop *Vandalia,* sailed for a point off the Savannah River, a rendezvous chosen in the hope of fooling the Confederates. When the rest of the expedition departed the following morning it took most of the day for the vessels to clear the harbor and take the places assigned to them in Du Pont's sailing orders—11 battleships in line abreast, followed by the transports in three columns (line ahead), with two battleships on each flank and two others bringing up the rear.

Well aware that it would be impossible to keep the Confederates from learning that something big was afoot, the Union high command did its best to keep the expedition's destination a secret. Even the ships' captains did not know where they were bound when they sailed. Their instructions were to follow the flagship, but they carried sealed orders to be opened in case they became separated from the convoy. While these elaborate precautions may have kept most northerners in the dark, they did not conceal the truth from some important southerners. Almost before the ships were clear of Hampton Roads the Confederate Secretary of War telegraphed from Richmond to the Governor of South Carolina, the general commanding the defenses of Charleston, and General Thomas F. Drayton, commanding the troops at Port Royal: "I have just received information which I consider entirely reliable that the enemy's expedition is intended for Port Royal."[1]

Early Friday, November 1, the pleasant weather hitherto enjoyed by the expedition gave way to a southeast gale. At noon the wind reached hurricane force and the flagship hoisted a signal bidding each ship to take care of herself. Some of the vessels, never designed to withstand rough weather, suffered severely before the storm ended three days later.

The *Governor,* a coastwise steamer being used as an Army transport, was the first to find herself in serious trouble. About 4 P.M. Friday her hog braces broke. (Hog braces were trusses used to keep a steamship from hogging, or breaking her back, when suspended between

1. *Official Records of the Union and Confederate Navies in the War of the Rebellion* (Washington, D. C.: Government Printing Office, 1894–1922), Ser. 1, 6: 306.

two waves or on top of one with her ends unsupported.) A short time later her smokestack was carried away just above the deck, leaving her almost unmanageable because the reduced draft made it impossible to raise more than a few pounds of steam, and this necessitated stopping the engines frequently to permit the pumps to be used for a while. As if things were not already bad enough the packing blew out of a cylinder head at 3 A.M. Saturday, leaving the ship utterly helpless until repairs were made. With more than 100 men bailing and operating hand pumps while the engine was being fixed, water gained so fast that most of the men on board gave up hope, but somehow she stayed afloat through the night.

About 10 A.M. Saturday the *Isaac Smith,* another transport, which had been kept from sinking only by jettisoning all but one of her nine guns, sighted the *Governor* with her flag flying upside down—a seagoing distress signal. For some time it seemed probable that the *Isaac Smith* could do no more for the *Governor* than to report her fate. Finally the *Isaac Smith*'s people succeeded in launching a small boat and getting a hawser to the *Governor* so that she could be taken in tow. This hawser either parted or was accidentally let go. Another one passed by the *Isaac Smith* soon parted, leaving the *Governor* helplessly adrift again. In midafternoon the sailing frigate *Sabine* discovered the *Isaac Smith*'s plight. The *Sabine* and the *Governor* both anchored and the *Sabine*'s chain was slacked until her stern was close enough to the *Governor*'s bow to permit about 30 men to be transferred to the frigate by means of a breeches buoy. Because this method was so slow the *Sabine* dropped nearer to the *Governor* to allow the latter's people to jump from one ship to the other. About 40 more men were saved this way before the risk that the heavily built *Sabine* would pound the *Governor* to pieces made it necessary to separate the vessels. At daybreak Sunday the *Sabine*'s small boats were launched, but they did not dare to get too near the wildly rolling steamer, so the men in the *Governor* had to jump overboard and be picked up from the water. The *Governor* finally sank, but the heroic efforts of the *Sabine*'s crew saved all except seven of the nearly 700 men who had been in the transport.

Friday afternoon the armed transport *Winfield Scott*, carrying part of a Pennsylvania regiment, sprang such a bad leak that by the following morning there was five feet of water in her hold. With the pumps clogged by floating cargo it became necessary to bail the ship by hand and to jettison nearly everything on board except the clothing the men were wearing—even the soldiers' muskets went by the board along with the vessel's cannon. When the gunboat *Bienville* came to the *Scott*'s help some of the transport's crew, including her chief engineer, deserted their posts and tried to board the gunboat. This sort of behavior on the part of seamen caused a panic among the troops, but the *Scott*'s captain managed to calm them. Soon afterward the transport *Vanderbilt* (formerly a passenger liner) took the half-sunken *Winfield Scott* in tow and brought her safely to Port Royal, but she never left there again.

Early Saturday morning the steam sloop *Mohican* came upon the transport *Peerless* in dire straits. After a small boat from the *Peerless* had been swamped a volunteer crew launched one of the *Mohican*'s boats which rescued the 26 men of the transport's crew just before she carried her cargo of horses to the bottom.

The steamers *Belvidere, Union,* and *Osceola,* all carrying army stores, but no troops, disappeared without trace.

When the storm ended late Sunday only one other vessel could be seen from the flagship's masthead, but thanks to their sealed orders the ships' captains knew where to go. By Monday night half of the expedition reached the rendezvous; Tuesday the rest of the vessels still afloat joined the others at an anchorage 10 miles off Port Royal bar, just within sight of the low lying shore.

Monday morning the Coast Survey steamer *Vixen,* protected by the gunboats *Ottawa, Seneca, Pembina, Curlew,* and *Pawnee,* began buoying Port Royal bar and Fishing Rip Shoal. This task was accomplished by 3 P.M. and all of the ships on hand except the *Wabash* crossed the bar on the next high tide. The *Wabash* drew so much water that she had to wait another day for an extra high tide and even then she cleared the bar by a scant two feet. (Anyone who has ever piloted a large vessel in such circumstances will appreciate what the captain went through, but, as he said, too much was at stake to hesitate.)

Early Tuesday morning the *Ottawa, Seneca, Curlew,* and *Isaac Smith* made enough of a demonstration against the Confederate works to draw their fire in order to permit an estimate to be made of their strength. At 11 A.M., after the *Wabash* had crossed the bar, a council of war was held in her cabin. This gathering broke up so late that no attack on the forts was made that day—the ships had not been together long enough for a night movement even to be considered. Although Wednesday was bright and fair, a strong southwest wind kept the fleet idle; the sea was not really rough, but Du Pont saw no reason not to wait for still more advantageous conditions.

As a result of Tuesday's reconnaissance Du Pont decided to move against the two forts simultaneously. At early dawn Thursday the flagship signalled in succession "go to breakfast," "form line of battle," and, at about 9 A.M., "prepare for action." According to Du Pont's orders the fleet formed two columns, with intervals of a little more than a vessel's length between ships in the same column and a space equal to a vessel's length between the columns.

While the Union ships' anchors were being weighed Commodore Josiah Tatnall, commanding a small Confederate flotilla of converted merchantmen, had his broad pennant dipped in salute to his old messmate Captain Du Pont. However, General Drayton offered no salute to his brother Percival, commanding the *Pocahontas,* when that ship, which had been delayed by the storm, joined the Union fleet.

Port Royal was defended by two earthworks—Fort Walker on a low bluff at Hilton Head, and Fort Beauregard two and a half miles to the north at Bay Point on St. Helena Island—mounting a total of five 6-inch rifles and 47 smoothbores ranging from 10-inch columbiads down to 12-pounders. As Stringham had done at Hatteras Inlet, Du Pont sought to offset the forts' supposed superiority over his fleet's total of 155 guns by keeping most of his ships constantly in motion.

The vessels of the starboard, or flanking column—the *Bienville, Seneca, Penguin,* and *Augusta*—concentrated their fire on Fort Beauregard as they steamed past it, then they anchored north of Hilton Head where they could enfilade Fort Walker while standing ready to deal with Tatnall's flotilla if it attempted to attack the transports, as Du Pont expected it would.

The ships of the port, or main column—the *Wabash, Susquehanna, Mohican, Seminole, Pawnee, Unadilla, Ottawa, Pembina,* and *Isaac Smith,* towing the *Vandalia*—opened fire as soon as they came within range of Fort Walker, successively using their bow, broadside, and stern guns until the *Vandalia* reached a point where her stern guns would no longer bear on the target; then the flagship turned to port and again led the way past the fort. Thus the main column described a series of ellipses with their shortest distances from the fort varying from 600 to 800 yards. Each round trip took a little more than 30 minutes with almost half of the time spent close aboard the fort.

At the completion of the second ellipse the ships of the main column drew off to let their crews eat a midday meal. Soon after action was resumed Fort Walker was abandoned in such haste that its guns were not spiked and stores and personal belongings were left undamaged. All of the ships now turned their attention to Fort Beauregard. Here the Confederates fought bravely until about 4 P.M., then they stampeded. (It should be remarked that both forts were garrisoned largely by militiamen who had stood up well under a heavy fire until this time.)

Although the ships were hit frequently and they maintained a brisk fire, particularly against Fort Walker, there were remarkably few casualties on either side during this engagement. The Confederates had 11 men killed, 48 wounded; the Union ships' losses were eight killed, 32 wounded in the battle, and one man hurt when he stepped on a booby trap in Fort Beauregard after the fight ended. (One reason the ships got off as easily as they did is that they stood in close with Fort Walker and this caused the Confederate gunners to fire so high they often missed their hulls.)

The loss of Port Royal was recognized by General Robert E. Lee of the Confederate Army as a severe blow to his country's cause. He wrote to the Secretary of War on November 9, 1861:

> The enemy having complete possession of the water and inland navigation [in the vicinity of Port Royal], commands all the islands on the coast [the so-called sea islands] and threatens both Savannah and Charleston, and can come in his boats, within 4 miles of this place [Lee's headquarters at Coosawatchie, South Carolina]. His sloops of war and large steamers can

> come up the Broad River to Mackay's Point, the mouth of the Pocotaglio [River], and his gunboats can ascend some distance up the Coosawatchie and Tulfinny [Rivers]. We have no guns that can resist their [ships'] batteries, and have no resources but to prepare to meet them in the field.[2]

Events at Port Royal so clearly demonstrated the power of shell guns in ships that the Confederates immediately abandoned all of their minor defensive positions along the coast except for a few that were difficult to approach by water. Paradoxically this led to a postponement of the move against Fernandina while the Union Army and Navy occupied, fortified, and garrisoned St. Helena, Beaufort, North Edisto, South Edisto, and Warsaw and Ossabaw Sounds.

During this interval the Confederates conceived the idea of making a counterattack on Beaufort and Port Royal Island (on which the town of Port Royal is located) and perhaps capturing a regiment or more of Yankees.

For this purpose the Confederates intended to place obstructions at Seabrook's Point on the Whale Branch two and a half miles from the Port Royal Ferry on one side of the Coosaw River and Boyd's Neck on the other side to prevent the ascent of gunboats; then heavy batteries were to be erected to protect the barrier and a large enough force thrown across to sweep away the Union troops on Port Royal Island.

General Thomas W. Sherman (sometimes called "the other Sherman"), commanding the Union troops on the South Carolina coast, learned of the Confederates' plans and called upon the Navy to help him thwart them.

Du Pont detailed two gunboats, an armed tug, and four boats armed with howitzers to go up the Beaufort

---

2. *Civil War Naval Chronology* (Washington, D. C.: Naval History Division, Office of the Chief of Naval Operations, Navy Department, n. d.), pt. 1, p. 34.

River to the Coosaw River to protect the troops while they landed and as they marched along the shore. Several other gunboats were simultaneously to move into the Broad River and the Whale Branch to bombard a battery supposed to have been erected opposite Seabrook's Point, then to attack the batteries at Port Royal Ferry.

These forces carried out their mission on January 1 and 2, 1862. Thereafter the Confederates left Port Royal strictly alone.

The long delayed Fernandina expedition, comprising 14 battleships and six transports, finally departed from Port Royal on February 28, 1862 and reached its destination March 2. This show of force frightened the Confederates so much they abandoned the place without a fight, although they could have held it at least for some time.

During the summer of 1861 two former United States Senators, James M. Mason of Virginia and John Slidell of Louisiana, were named by the Confederate Government as "Ministers Plenipotentiary" to Great Britain and France respectively.

The fact that these appointments had been made quickly became known to the Union authorities, who probably urged extra vigilance upon the officers of the blockading squadrons in an effort to keep the pair from leaving the Confederate States. Nevertheless, they sailed from Charleston on Saturday, October 12, 1861, in the S.S. *Theodora* bound either for Nassau, the Bahamas, or Cardenas, Cuba.

The *Theodora* had been chartered by the Confederate Navy to keep a daily watch on the Union blockaders at Charleston. Her draught was so shallow that she did not have to stay in the channel and she was so fast the Union vessels finally quit chasing her, especially as she had never tried to get past them. Thus she was easily able to slip out of port the one time she really wanted to do so.

When the *Theodora* arrived at Cardenas on Monday, October 14, Mason and Slidell learned that there was no regular means of travel between there and St. Thomas, where they intended to embark in one of the British Royal Mail Steamship Company's vessels to cross the Atlantic. Therefore, they went to Havana, a port of call on the British line.

Soon after Mason and Slidell reached Cuba the U.S.S. *San Jacinto,* commanded by Captain Charles Wilkes, put into Havana where Wilkes hoped to obtain some news about the Confederate cruiser *Sumter* for which he was searching. He got no information about the *Sumter,* but he did learn that the Confederate ministers had booked passage to England in the Royal Mail Steamship Company's *Trent,* scheduled to call at St. Thomas on November 7. Wilkes quickly decided to regard the *Trent* as a blockade runner on the grounds that she would be carrying contraband in the persons of Mason and Slidell.

Wilkes left Havana and put into Key West, Florida, where he hoped to find another Union ship to assist him in his self-appointed mission. Perhaps he also hoped to find somebody to share the responsibility for the action he planned to take for he certainly cannot have expected it to be dangerous enough or difficult enough really to have foreseen any need for help. In any case there was no other Union vessel to be found near Florida so he had to carry on alone.

On November 8 the *San Jacinto* reached a position in a narrow part of the Bahama Channel some 240 miles from Havana, about nine miles northeast of Paredon del Grande lighthouse, where Wilkes had decided to lie in wait for the *Trent.* When the liner appeared toward noon that day the *San Jacinto* fired a shot across her bow.

Hoisting British colors, the *Trent* held her course and speed until a shell was fired past her bow. At this grim warning she hove to, but her captain haughtily refused to show his ship's papers or her passenger list. However, when Mason and Slidell heard their names mentioned by the boarding officer they made themselves known. They were asked to return to the *San Jacinto,* but they refused to leave the *Trent* until they were compelled to do so by an officially "overwhelming force." Actually they were merely touched on the shoulders by two men, but for the record they had been forcibly removed from a British vessel on the high seas. Having legally established this record they went willingly, even happily, on board the *San Jacinto.*

Almost from the beginning of the war Great Britain and the United States had been feuding diplomatically, because Great Britain believed her rights as a neutral were being violated by the blockading squadrons while the United States felt that British "neutrality" was in fact highly benevolent toward the Confederate States. In these circumstances the Confederates logically hoped the *Trent* affair would gain a valuable ally for them. It certainly came near doing so. The usual practice in connection with an event such as Wilkes's act was for the injured party to make representations based on the assumption that the thing was a subordinate's error which could be disavowed and for which an apology and, if need be, reparation could be made. Instead of following this course the British Cabinet prepared a hot-tempered note for transmission to Washington in which it was forcefully asserted that Great Britain's neutral rights had been violated, a shrill demand was made that Wilkes's action be disawoved, and war was virtually threatened unless Mason and Slidell were immediately set free. Queen Victoria's consort, Prince Albert (who was suffering from the illness of which he died two weeks later), caused the intemperate language of this note to be moderated considerably before it was delivered.[3]

Most northerners wildly applauded Wilkes's exploit, no doubt to his gratification for he was an egotistical man. Congress voted him a medal; he was wined and dined in Boston where he landed Mason and Slidell to be incarcerated in Fort Warren, a coast defense installation; and Gideon Welles officially congratulated him on the "public service [he had] performed in the capture of the rebel commissioners Mason and Slidell."[4]

Welles changed his tune when he stopped to think about the explosive situation Wilkes had created by not bringing the *Trent* into port with Mason and Slidell in her. If he had done so the case would have been handled by an admiralty court as a possible violation of neutrality. The removal of the pair from the ship made it a political matter and it was easy to foresee that there would be a strong British reaction.

Actually the decision not to seize the *Trent* but only to remove Mason and Slidell was made by the *San Jacinto*'s executive officer, who led the boarding party, because he thought that taking her would create a grave danger of a British-American war. However, Wilkes, who could have overruled the executive's decision, said in his report that he had not sent the ship into Key West because he could ill spare enough officers and men for a prize crew and he did not want to inconvenience her passengers.

Fortunately for the Union, President Lincoln and Secretary of State Seward kept their heads. They realized that no matter how much Wilkes's seizure of Mason and Slidell had delighted many northerners it had flagrantly disregarded the principle of the inviolability of neutral ships, the defense of which had been one of the causes

3. See Hector Bolitho, *Albert Prince Consort* (Indianapolis: The Bobbs-Merrill Company, Inc., 1964), pp. 224–25, for relevant passages from the British note as originally drafted and as modified by Prince Albert.

4. *ORN*, Ser. 1, 1: 148.

of the War of 1812. Seward wrote to the American minister to Great Britain that Wilkes had acted on his own initiative, entirely without orders from his superiors. The fact that this was said before any British communication on the subject had reached the United States eased the crisis. It ended altogether when the Federal Government, after receiving the comparatively mild British note, agreed to release Mason and Slidell. However, Seward took occasion while informing the British government that the pair would be cheerfully liberated to point out that by condemning Wilkes's action Her Majesty's government had practically acknowledged that the erstwhile British practice of seizing alleged Englishmen found in American vessels had always been wrong under international law.

Without conceding Seward's claim in this matter the British Foreign Secretary accepted the statement that Wilkes's action was not authorized and the release of Mason and Slidell as constituting such reparation as Great Britain had a right to expect. However, the general tone of the British press, particularly of the part supposed specially to speak for the government, remained distinctly unfriendly to the United States. English newspapers frequently referred to the benefits to be gained by recognizing the Confederate States diplomatically and breaking the blockade. Suggestions were even made for an armed intervention similar to that of the allied powers (Great Britain, France, and Russia) in the war between Greece and Turkey in 1821 to 1829.

The *Trent* affair had a comic sequel; some British troopships en route to Canada in case of war with the United States found the St. Lawrence River frozen and the soldiers were set to the destination by rail, via Portland, Maine.

On October 12, 1861 (the day Mason and Slidell left Charleston and the first of the Eads gunboats was launched) an event of no great military importance but of considerable historic interest occurred in a two-mile-wide roadstead known as the Head of the Passes, a place about 15 miles from the Gulf of Mexico where the Mississippi River divides into four branches called Pass a l'Outre, Southeast Pass, South Pass, and Southwest Pass.

For some time past the 14-gun steam sloop *Richmond,* the three-gun side-wheeler *Water Witch,* the 20-gun sailing sloop *Vincennes,* and the 16-gun sloop *Preble* had been stationed at the Head of the Passes to blockade New Orleans. About 4 A.M. a strange craft was discovered close aboard the *Richmond,* which was taking coal from the schooner alongside. Before an alarm was raised the stranger rammed the *Richmond* hard enough to tear the schooner loose and punch a small hole in the sloop below the water line. The attacking vessel immediately dropped astern of the *Richmond,* remained there briefly, then steamed slowly upstream entirely undamaged by hits made on her at point-blank range by the *Vincennes* and *Preble.* When three fire rafts came drifting down the river a few minutes later the *Richmond, Vincennes,* and *Preble* fled toward Southwest Pass in such haste that the first two of them ran aground. Although the fire rafts came nowhere near either of the stranded vessels, the captain of the *Vincennes* made preparations for her destruction and abandoned her. One sailor, who used more judgment than his commanding officer had, cut the fuse leading to the magazine before he left the ship, thus preventing her from being blown up. When the captain was ordered to return to his ship he threw most of her heavy guns and stores overboard to get her afloat, then hastened after the *Richmond* and *Preble* with the *Water Witch* covering their flight.

The Confederate vessel that so ignominiously routed the Union ships, the *Manassas,* is historically important because she was the first ironclad that ever engaged an enemy warship.

The *Manassas* was originally a tugboat named the *Enoch Train,* built in Medford, Massachusetts, in 1855. She was converted at the expense of a syndicate of New Orleans businessmen into a cigar-shaped ram 128 feet long with a turtleback deck made of 12 inches of timber covered with 1½ inches of iron. She carried a 9-inch Dahlgren gun which protruded through a forward hatch so small the piece could not be trained but had to be aimed by steering the vessel. The *Manassas* was never used as a privateer, but was commandeered by the Confederate Navy.

At the time of her attack on the *Richmond* the *Manassas*'s one gun was not in serviceable condition and her engine was in such poor repair that she could not make more than six knots in slack water. When she rammed the *Richmond* the *Manassas* suffered more damage than she inflicted. The collision wrenched her prow badly, carried away her smokestack, and put a condenser out of order. In these circumstances she was compelled to retire from the scene after she struck the *Richmond.*

Probably the curvature of the *Manassas*'s deck caused projectiles to glance off, and thus afforded her more protection than her light armor did, but the significant fact that she had not been even slightly damaged by heavy gunfire at short range apparently went entirely unnoticed by everybody connected with the Union Navy Department. It is true that on this occasion the Department was much disturbed about the incompetent and disgraceful behavior of most of the ships' captains involved in the affair. They had learned from escaped slaves of the existence of an ironclad at New Orleans, but they had taken no precautions against being attacked by it (there was not even a picket boat on duty when the *Manassas* came down the river) and the only vessel that even offered to fight instead of immediately running away was the little *Water Witch*. However, hardly anybody in the Union Navy Department took much interest in ironclads until one of these ships demonstrated her destructive capabilities in a way that simply could not be ignored.

In an effort to prevent the Union Army and Navy from exploiting the foothold gained at Hatteras Inlet the Confederates strongly fortified Roanoke Island which commanded the southern entrance to Albemarle Sound and the "back door" to Norfolk.

This island, 12 miles long and three wide, lies between Croatan Sound on the west and Roanoke Sound on the east. The only road running its length was guarded by a battery of three fieldpieces located just north of a causeway across a swamp as wide as the island. These guns could sweep a broad area and felled trees and other obstructions were placed in front of them to impede an attacking force.

Because of Roanoke Sound's shallowness, its eastern side was defended by only a few guns in Battery Ellis. However, there were three strongholds on the side facing Croatan Sound. The southernmost of them, Fort Bartow, was an earthwork mounting eight 32-pound smoothbores and a 68-pound rifle, with three fieldpieces protecting its rear. A mile and a half to the north Fort Blanchard contained four 32-pounders. Fort Huger, another mile farther up the island, had twelve 32-pound rifles and smoothbores in it. On the mainland, opposite Fort Huger, stood Fort Forrest containing seven 32-pounders. A double row of stakes and some sunken vessels stretched across Croatan Sound between Forts Huger and Forrest. Above this barrier there was a squadron of eight gunboats, each of which mounted a 32-pound smoothbore and one also had a 30-pound rifle. The 4000 officers and men of the land forces and the gunboats were under orders to fight to the last ditch if they were attacked.

In the fall of 1861, soon after General George B. McClellan was named General in Chief of the Union Armies, he proposed that a joint expedition be sent against Roanoke Island. At this time he was planning his peninsula campaign and he realized that a Union division based on the island and able to operate in North Carolina would pose a serious threat to the Confederate lines of communication and thus ease the pressure on his army. Goldsborough, who could see that possession of the island would tighten the blockade of North Carolina, welcomed McClellan's proposal and promised his hearty cooperation.

The 12,000 troops assigned to the expedition boarded 46 transports at the Naval Academy at Annapolis on January 5, 1862, the day after a storm had dropped two or three inches of snow on the ground. At Hampton Roads the transports were joined by 19 shallow draught gunboats and the entire force arrived at Hatteras Inlet on January 13, just as a severe northeast storm began.

Before the weather moderated much havoc had been done to some of the ships. The transport *Whitehall* was so badly damaged she had to be sent back to Hampton Roads for repairs. The gunboat *Zouave* sank as a result of pounding on her own anchor after she had passed through Hatteras Inlet. The transport *Pocahontas* was wrecked on the outside of Hatteras Island and most of the 113 horses she had on board were lost. The transport *City of New York* went ashore on the outer beach of the island and the rest of the fleet either failed to recognize her plight or ignored her distress signals. Finally four men managed to row ashore and help was sent to their shipmates who had been clinging to the vessel's bulwarks and rigging for nearly three days. Her cargo—consisting in part of 400 barrels of gunpowder, 1500 rifles, 200 artillery shells, hand grenades, and tents —was a complete loss.

After the storm ended there was a further delay because many of the transports drew more water than their owners in their eagerness to charter them to the War Department had admitted, and there was less water over the bar than Goldsborough had expected to find. In order to create a channel of the necessary depth (eight feet) some of the larger steamers were deliberately run onto the bar and held there by means of anchors carried out ahead of them by small boats; as the tide ebbed the current washed sand from under them until they floated again, then the process was repeated nearer Pamlico Sound. This method of dredging, though highly ingenious, was so slow that it took until February 4 to get the last vessels through the inlet into the sound.

Early the following morning the fleet started for Roanoke Island in three columns, headed by the gunboats. The channel buoys had been removed by the Confederates, whose pilots could find their way without their aid and, with leadsmen constantly sounding, the fleet groped its way to Stumpy Point, 10 miles below Fort Blanchard where it anchored at sunset because it was simply impossible for it to go on in the dark. In the morning the ships got under way in a drizzle that soon gave way to a fog so dense they had to anchor again.

By this time the Confederates had learned that a number of "Lincoln gunboats" were making their way up Pamlico Sound so the C.S.S. *Curlew* was sent down from Roanoke Island to reconnoiter. The Union vessels made no attempt to molest the *Curlew* because Captain Goldsborough hoped her report of the great strength of the expedition would overawe the Confederates and perhaps save him from the necessity of fighting them. Since the Confederates were less easily frightened than Goldsborough hoped they would be this early example of psychological warfare failed to effect its purpose.

Favored at last by bright, clear weather the Union force closed the southern end of Roanoke Island about 9 A.M., February 7. Three gunboats steamed ahead toward Sandy Point near Ashby's Harbor, a cove about two miles below Fort Bartow, where the troops were to land, while the other gunboats gathered around the flagship. The Confederates quickly made it clear by their heavy gunfire that they intended to hold the island if they could.

Six hours later the troops began landing on Sandy Point in a manner anticipating vaguely similar operations in World War II. A number of steamers each towed 20 boats filled with soldiers toward the chosen landing place; when enough speed was reached to carry the boats ashore they were cast off to run for the beach. However, the shore sloped so gradually that the men had to disembark into icy water waist deep and wade ashore under fire from a considerable body of Confederate infantry and several pieces of field artillery. But as soon as the gunboat *Delaware* was maneuvered into a position from which she could enfilade Ashby's Harbor, shrapnel from her carefully aimed 9-inch guns dispersed the Confederates. By midnight 10,000 soldiers and the crews of six Navy launches, with as many boat howitzers, had landed.

In the morning the troops moved up the island in three columns. The center column, accompanied by the Navy howitzers and their crews, attacked the Confederate fieldpieces while the other columns made their way through the swamp and flanked the battery, thus opening the road to Fort Bartow.

Not having any means of communicating with the troops who were out of his sight (such as radio affords today) Goldsborough literally had to "play it by ear." At nine o'clock, when the sound of gunfire indicated that the troops were hotly engaged, the gunboats moved past Ashby's Harbor and closed Fort Bartow. Four hours later, again judging by the noise on shore that the Union troops had gained the rear of the fort, Goldsborough ordered the gunboats to cease fire and set some of them to work breaking through the channel obstructions. By 5 P.M. this task had been accomplished, the forts on the island had been abandoned, and the Union troops were mopping up.

This operation affords an excellent example of the value of interservice cooperation. The naval bombardment was not enough by itself to overcome the forts, but it contributed materially to the Union victory. And, of course, the troops could not have been transported to the island by water if it had not been for the Union's overwhelming seapower.

The Union Navy's casualties in this affair were six killed, 17 wounded, two missing. The Confederates, fighting in prepared positions, suffered fewer casualties than the Union Army did, but nearly 3000 men were captured with their small arms, several pieces of field artillery, and all of their big guns.

The day after Roanoke Island was taken all of the ammunition left in the Union ships was loaded into 12 gunboats and they set out in pursuit of the Confederate steamers which had retreated to Elizabeth City at the mouth of the Pasquotauk River. They sailed the following morning under strict orders to do no shooting except on orders from the flagship. About 8 A.M. they closed the Confederate flotilla lying between Fort Cobb, a battery of four 32-pounders on Cobb's Point, and the schooner *Black Warrior,* armed with two similar pieces. The Confederates opened fire at long range; the Union vessels made no reply until the range had been reduced to three-quarters of a mile when the flagship signalled: "Dash at the enemy."[5]

5. Daniel Ammen, *The Atlantic Coast* (New York: Charles Scribner's Sons, 1883), p. 184.

Passing between Fort Cobb and the *Black Warrior* at full speed the U.S.S. *Commodore* rammed and sank the Confederate flagship *Sea Bird;* the U.S.S. *Delaware* forced the C.S.S. *Fanny* ashore where she was blown up by her own people; the C.S.S. *Black Warrior*'s crew ran her ashore and set her on fire; the C.S.S. *Appomatox* tried to escape into the Dismal Swamp Canal, but she drew too much water so she was destroyed by her crew; the U.S.S. *Whitehead* and *Valley City* compelled the Cobb's Point battery to surrender; and boarders from the U.S.S. *Ceres* carried the C.S.S. *Ellis,* thus ending one of the most spirited little engagements of the Civil War. (Incidentally, the *Ellis* was one of the last vessels ever captured by a boarding party.)[6]

During this melee a shell passed through the *Valley City*'s magazine and exploded in a locker containing some rockets. When the vessel's captain went to examine the damage he found a gunner sitting with great aplomb on an open barrel to keep the fire from reaching the gunpowder it contained. He was awarded a Congressional Medal of Honor for his deed.

6. For a description of boarding see Howard P. Nash, Jr., "Boarders Away," *Tradition* 4, no. 8 (August 1961): 15 ff.

4

First Battle of Ironclads

For several years before the Civil War began naval authorities in Europe and America had been considering the potential value of ironclad ships. By 1861 the French Navy had seven such vessels, the British Navy four, and the United States Navy none. Because no such craft had ever been tested in battle different opinions about their worth were held by various persons. In America almost nobody connected with the Union Navy Department thought they would be of any great value, but the farsighted Secretary of the Confederate Navy wanted some of them.

As early as May 8, 1861, Mallory urged the Confederate Congress to appropriate funds for the construction of one or more ironclads. "Such a vessel," he said, "at this time could traverse the entire coast of the United States, prevent all blockade, and encounter with a fair prospect of success their entire navy."[1]

Soon after the money Mallory sought was made available Lieutenant John M. Brooke and Naval Constructor

---

1. *Official Records of the Union and Confederate Navies in the War of the Rebellion* (Washington, D. C.: Government Printing Office, 1894–1922), Ser. 2, 2: 69. (Hereafter cited as *ORN*.)

John L. Porter (both formerly of the United States Navy) submitted plans for and a model of an ironclad ship. By coincidence their ideas were strikingly similar. The chief difference between them was that Brooke envisioned a seagoing vessel, Porter one intended only for harbor defense. Mallory, who liked their basic ideas, ordered them to work out a design in collaboration with William P. Williamson, Chief Engineer of the Confederate Navy. By this time the *Merrimack* had been raised and at one of the trio's conferences Williamson suggested that time and trouble could be saved by using her engines in the ship they were designing. (The *Merrimack*'s engines had been condemned and were to have been replaced. They had, of course, deteriorated further as a result of having been under water for a month. However, they could be reconditioned quicker and more easily than new ones could be built in the Confederate States, where there was a grave lack of skilled laborers.) The idea of using the *Merrimack*'s engines led them also to decide to use her hull which had been only slightly damaged by fire because she had sunk so quickly.[2]

The vessel Brooke, Porter, and Williamson designed was named the *Virginia,* but she is often miscalled the C.S.S. *Merrimack* (a vessel that never existed).

As Mallory said, "The *Virginia* was a novelty in naval architecture, wholly unlike any ship that ever floated" before she was launched.[3]

She had a casemate 174 feet long, 7 feet high, with walls 28 inches thick (24 inches of wood covered with two 2-inch layers of iron), sloping inward at an angle of 35 degrees to deflect projectiles hitting them. Both ends

2. F. Moore, *Rebellion Record* (New York: G. P. Putnam; D. Van Nostrand, 1861–67), vol. 3, docs., 279–80; Robert U. Johnson and Clarence Buel (eds.) *Battles and Leaders of the Civil War* (New York: The Century Co., 1884–87), 1: 715–17. (Hereafter cited as *BL.*)

3. *Civil War Naval Chronology* (Washington, D. C.: Naval History Division, Office of the Chief of Naval Operations, Navy Department, n. d.), pt. 2, p. 29.

of the casemate were rounded to permit two pivot guns (7-inch Brooke rifles) to be used as bow and stern chasers or in broadside. Her broadside battery included two 6-inch Brooke rifles and six 9-inch Dahlgrens. A cast iron ram several feet long was bolted to her bow a couple of feet below the water line. In cruising trim she had a free-board of 3½ feet; on preparing for action she took in enough water ballast to bring her deck awash. Thus, as her surgeon said, when she was ready for battle she "bore some resemblance to a huge terrapin with a large chimney about the middle of its back."[4]

Many southerners, including Mallory who should have known better, fondly hoped the *Virginia* could and would quickly win the war for the Confederate States by raising the blockade at Hampton Roads, then attacking Washington, New York, and other Atlantic coast ports. Actually the *Virginia*'s draught was so great (22 feet aft) that even if she had met with no opposition she could not have made her way up the Potomac River to Washington and she turned out to be so unseaworthy she would quickly have sunk if she had ever ventured outside of Hampton Roads. But, as she was to demonstrate, she was more than a match for any wooden vessel possessed by the Union Navy. To that extent she did fulfill Mallory's hopes.

The Union Navy Department gained a great deal of information about the projected Confederate ironclad from spies, deserters, and southern newspapers. However, the Department paid little heed to what it learned and nobody in Washington thought for a moment of trying to obtain a vessel specially designed to contend with the *Virginia*. One of the few northerners who was at all concerned about the work being done on the *Merrimack* was Charles Ellet, the designer of the Union Army

4. *BL*, 1: 718.

rams, and he was perturbed not because the Confederate craft was to be armored, but because the Union Navy had no vessels able to withstand an attack by a well-built ram.[5]

However, as a result of a chain of purely fortuitous circumstances at the eleventh hour the Union Navy acquired a ship capable of coping with the *Virginia.* And as a result of additional luck the Union vessel happened to be at the right place at the right time.

A report concerning existing French and British armored ships submitted by Commander Dahlgren led the Secretary of the Navy to say to the Thirty-seventh Congress when it convened in special session on July 4, 1861:

> Much attention has been given within the last few years to the subject of floating batteries, or iron-clad steamers. . . . The [present] period is, perhaps, not one best adapted to heavy expenditures by way of experiment, and the time and attention of some of those who are most competent to investigate and form correct conclusions on this subject are otherwise employed. I would, however, recommend the appointment of a proper and competent board to inquire into and report in regard to a measure so important; and it is for Congress to decide whether, on a favorable report, they will order one or more iron-clad steamers, or floating batteries, to be constructed, with a view to perfect protection from the effects of present ordnance at short range, and to make an appropriation for the purpose.[6]

A month later the Congress, as little excited about ironclads as Welles was, authorized the appointment of the board he had mentioned and appropriated $1.5 million to be spent for armored ships at the Secretary's discretion. (The bill passed the Senate by 18 votes to 16. This vote certainly does not indicate much interest in or concern about the need for ironclads.)

In due time (September 16) the ironclad board rec-

5. Ibid., 1: 453–54.
6. Ibid., 1: 616.

ommended the acceptance of three of the 17 proposals it had considered. Soon afterward the Navy Department placed contracts for the construction of the *Galena, New Ironsides,* and *Monitor.*

Except that they were to be armored there was nothing remarkable about the first two of these vessels. The *Galena's* hull had an unusual degree of tumble home, otherwise she resembled a typical gunboat of the period. Rigged as a topsail schooner, equipped with two steam engines driving one propeller, she had a thin iron skin over closely spaced iron bars four inches thick. The *New Ironsides* (so named in honor of *Old Ironsides,* the U.S.S. *Constitution*) was similar to the ironclads then being built in Europe. She had wooden ends with armor only along her sides. Mounting 18 guns, including a pair of 200-pound rifles, she was launched as a bark-rigged steamer, but before she saw any action her heavy masts were replaced by poles. However, the *Monitor,* designed by and built under the supervision of a Swedish-born inventor named John Ericsson, was something as novel as the *Virginia.*

The *Monitor* had a raftlike deck 172 feet long and 41 feet wide fastened by a single set of rivets to a scow-shaped hull 122 feet long by 34 feet wide. Near the bow there was a rectangular pilothouse, standing about 4 feet high above the deck and extending a couple of feet below it, made of 9 by 12 inch iron bars, mortised and tenoned, and fastened with iron bolts at the corners. Instead of windows the pilothouse had 1-inch slits on all sides. Amidships there was a revolving turret containing two 11-inch Dahlgrens. This structure, 9 feet high with an inside diameter of 20 feet, had walls 8 inches thick, made of 1-inch plates held together with 1¼ inch bolts. Abaft the turret there were two telescoping smokestacks side by side and some intake pipes leading to belt-driven blowers used to ventilate the ship's interior. The deck was made of iron 2 inches thick and the laminated side armor, varying from 5 inches thick above the water line to 3 inches below it, was backed with 27 inches of wood. An anchor well in the forward overhang, with a hand-operated windlass in a compartment immediately behind it, made it possible for a four-pronged anchor to be dropped or drawn up into the well without anyone having to expose himself on deck. (A defect of this ingenious arrangement was that the hawsehole leading from the anchor well into the windlass room was only a few inches above the normal water line, making it necessary to drive a rope gasket around the chain to prevent leakage. Neglect of this precaution caused one monitor to sink at her mooring on a fairly calm day.)

Because the *Monitor* had a freeboard of only 24 inches forward and 18 inches aft her engines, boilers, magazines, staterooms, berth deck, etc., were all below the water line. To permit openings to be made in the hull in order to get rid of waste products Ericsson designed an ingenious contrivance with an air pump attached to it. As Ericsson's secretary and biographer described this device: "Waste matter was dropped into a pipe closed at the lower end. The upper end of the pipe was then shut, the lower end opened in its turn and the force-pump turned on, driving out the water with its contents."[7]

On one occasion an officer using a water closet of this sort "omitted an essential part of this ceremonial [and] found himself suddenly projected into the air at the end of a column of water [with a head of almost 10 feet] rushing up from the depths of the ocean,"[8] like a celluloid ball in a shooting gallery.

There is a persistent but wholly unfounded legend to the effect that Ericsson built the *Monitor* at the urgent

7. William Conant Church, *The Life and Times of John Ericsson* (New York: Charles Scribner's Sons, 1911), 1: 261.
8. Ibid.

request of the Union Navy Department and rushed her to completion barely in time to meet the *Virginia.* Actually the Department heard about Ericsson's ship by sheer chance and he built her in less than five months simply because he was in the habit of working fast.[9] (In the 1840s he built a 770-ton steamer in six months and it took him less than a year—April 1852-January 1853—to build the 2200-ton caloric ship *Ericsson.*)[10]

Ericsson designed the U.S.S. *Princeton,* the United States Navy's first screw-propelled warship, and one of her 12-inch, 225-pound pivot guns, in the 1840s. Her other big gun, an imperfect copy of Ericsson's piece, was designed by and made under the supervision of Captain Robert F. Stockton, the *Princeton*'s commanding officer. On February 17, 1844, Stockton's gun burst and killed the Secretary of the Navy and several other bigwigs who had been invited on a junket in the new ship. Stockton managed to shift the entire blame for the gun's failure onto Ericsson's shoulders. Even worse, as a result of Stockton's machinations Ericsson had to get a judgment from the United States Court of Claims for the money owed him for his work on the *Princeton* and then he had great difficulty in collecting it.[11] In these circumstances he swore he would never again have any dealing with the Navy. He changed his mind in 1861 as a result of a friendly trick played upon him by C. S. Bushnell, the designer and builder of the *Galena.*

When Bushnell showed plans of the *Galena* to the ironclad board some doubts were expressed about the ship's ability to float with the great weight of armor and armament she would have to carry. Bushnell consulted Ericsson in the matter. After reassuring Bushnell that

9. Howard P. Nash, Jr., "A Civil War Legend Examined," *American Neptune* 23 (July 1963): 197 ff.
10. Church, *John Ericsson,* 1: 189, 237.
11. Howard P. Nash, Jr., "The *Princeton* Explosion," *American History Illustrated* 4, no. 5 (August 1969): 4 ff.

the *Galena* was soundly designed Ericsson showed him a cardboard model, made several years earlier, of an ironclad. Bushnell urged Ericsson to submit the model to the ironclad board. Ericsson flatly refused to do this, but Bushnell won his reluctant permission to show it to Welles. Welles liked it a great deal, so did Fox, and so did the President; the ironclad board laughed at it and told Bushnell to "take the little thing home and worship it, as it would not be idolatry, because it was made in the image of nothing in the heavens above or on the earth below or in the waters under the earth."[12]

Bushnell, convinced of the merits of Ericsson's design, shrewdly led him to believe the board had liked it, but wanted him to explain a few details about it. Thus deluded, Ericsson appeared before the board only to be told that his idea had been scornfully rejected. As he was turning angrily away he was told that the reason for the board's attitude was its belief that the proposed vessel would lack stability. This statement roused his fierce professional pride and he defended his design so eloquently that he won over the board.

At Welles's request Ericsson began building the *Monitor* without waiting for a formal contract. However, the ironclad board had some second thoughts after Ericsson's departure so the contract sent to him three weeks later provided that if the vessel he built should fail "in the security or successful working of the turret, or in her buoyancy to float and carry her battery," or to make nine statute miles an hour all of the money advanced by the Navy Department toward her construction would have to be refunded.[13]

These provisions angered Ericsson to such an extent that he almost stopped work on the *Monitor* when he received the contract. He completed her only because he

12. *BL,* 1: 749.
13. Fortieth Congress, Second Session, Sen. Exec. Doc. No. 86, pp. 5, 7.

was unwilling to have his financial backers, including Bushnell, lose the money they had advanced toward her construction.[14]

By the time the *Monitor*'s keel was laid on October 25, 1861, the reconstruction of the *Merrimack* was well under way, but this fact did not disturb anybody connected with the Union Navy or particularly interest Ericsson. The naval officers who gave the matter any thought were perfectly confident that the Confederate ironclad could not be dangerous to anything so formidable as a frigate, whatever she might be able to do to smaller vessels. The fact that Ericsson did not design his ship particularly to cope with the *Virginia,* despite statements he made later,[15] can be inferred from a letter he wrote to the Assistant Secretary of the Navy on January 20, 1862, saying:

> In accordance with your request, I now submit a name for the floating battery [which was launched that day] at Greenpoint. The impregnable and aggressive character of this structure will admonish the leaders of the Southern Rebellion that the batteries on the banks of their rivers will no longer present barriers to the entry of the Union forces. The iron-clad intruder will thus prove a severe monitor to those leaders. But there are other leaders who will also be startled and admonished by the booming of the guns from the impregnable iron turret. "Downing Street" will hardly view with indifference this "Yankee Notion," this monitor. To the Lords of the Admiralty the new craft will be a monitor, suggesting doubts as to propriety of completing those four steel clad ships at half a million apiece. On these and many similar grounds I propose to name the new battery *Monitor.*[16]

---

14. Church, *John Ericsson,* 1: 253.
15. See e.g., John Ericsson, "The Monitors," *The Century Magazine,* 31, new series 9 (December 1885): 282 ff.; reprinted under the title "The Building of the 'Monitor,'" in *Battles and Leaders of the Civil War,* 1: 73 ff.
16. Church, *John Ericsson,* 1: 254, ftn.

Since the Union Navy's existing ships had already demonstrated their ability to overcome Confederate batteries such as those at Hatteras Inlet and Port Royal the Navy Department did not regard the *Monitor* as likely to be a particularly valuable addition to the fleet.

While the *Monitor* was under construction the Navy Department was uncertain what use to make of her. "She was," as a contemporary put it, "a naval curiosity, and looked upon as an experiment on a small scale, which might work some changes in naval architecture, nothing more."[17]

At one time Fox thought of her in connection with the *Virginia*.[18] Later either he or Welles decided that if she were completed soon enough she could be used to better advantage as part of a New Orleans expedition then being planned,[19] of which more will be said in another chapter. She was not ready in time for that service so she was ordered to go to Hampton Roads and report her arrival there to the Navy Department.[20]

The *Monitor* departed from the Brooklyn navy yard at noon, Thursday, March 6, 1861, in tow of the *Seth Low,* a small side-wheeler. Soon after she sailed a telegram, signed by the Secretary of the Navy, was delivered to the yard ordering the *Monitor* to proceed directly to Washington without stopping at Hampton Roads.[21] The Navy Department had finally decided to assign her to the Potomac Flotilla to help to get rid of some annoying Confederate batteries on the Virginia bank of the Potomac River.[22] A tugboat detailed to relay Welles's message to the *Monitor* was unable to overtake her so a

17. P. C. Headley, *Great Rebellion* (Hartford: American Publishing Company, 1866), 1: 251.
18. *ORN,* ser. 1, 6: 538.
19. Church, *John Ericsson,* 1: 278.
20. *ORN,* ser. 1, 6: 659.
21. Ibid., ser. 1, 6: 691.
22. James Russell Soley, *The Blockade and the Cruisers* (New York: Charles Scribner's Sons, 1883), p. 67.

duplicate order was sent to Hampton Roads to be delivered on her arrival there.[23]

The weather, ideal when the *Monitor* left Brooklyn, changed during the night. By Friday noon a strong northwest wind was blowing and water was sweeping against the *Monitor's* pilothouse where it burst through the peepholes with force enough to knock the helmsman away from the steering wheel several times. In a little while water entering under the bottom of the turret began to fill the hull. Toward 4 P.M. seas breaking over the air intake pipes caused the belts used to drive the ventilator fans to become wet enough to slip badly, with the result that the engine room and stokehold quickly became so full of coal gas that the engineers and firemen were knocked unconscious. (Ericsson, who never admitted the existence of the slightest fault in connection with anything he ever designed or built, said in 1885 that there would have been no trouble in the pilothouse if the 5/8 inch peepholes he had specified had not been enlarged to 1 inch; he simply disregarded the fact that the ventilating system had not worked well at all.)[24]

While the firemen and engineers were being revived on top of the turret, the only place in the ship where they could get fresh air without being washed overboard, the untended fires almost went out. As the boiler pressure dropped the steam pumps stopped working. When the hand pumps were manned it was found that they could not lift enough water to be of any use. In the only hope of keeping the *Monitor* afloat the *Seth Low* headed directly toward land, almost at a right angle from the course toward Hampton Roads. Five hours' steaming in this direction brought the *Monitor* into calm water, but the respite thus gained was brief. Before midnight the sea grew dangerously rough again, and now, while the

23. *ORN,* ser. 1, 6: 691.
24. Ericsson, "The Monitors," 282.

tired crew struggled to keep the ship from sinking, the wire cables running from the steering wheel to the rudder jumped off their guidewheels. The *Monitor*'s steam whistle had not been installed so her people tried to hail the tug, but could not make themselves heard over the howling of the wind and, unfortunately, no arrangements had been made for visual signalling at night. With the *Monitor* pitching and yawing so wildly that only a miracle kept the towline from snapping the steering gear was repaired somehow. As soon as daylight made communication possible the *Seth Low* was ordered to stand in toward shore again. By 8 A.M. the *Monitor*'s weary crew could at last consider themselves safe from the elements. Thirteen hours later the *Monitor* reached Hampton Roads.

At noon, March 8, while the *Monitor* was somewhere south of Cape May, New Jersey, the *Virginia* left her pier in the Elizabeth River for the first time. None of her guns had ever been fired, her engines had been turned over only a few times, and there had been too many workmen on board for the crew to have done much drilling. Thus the ship was wholly untried and her officers and men were virtually strangers to each other. In view of these facts most of her people naturally supposed she was bound for a short trial trip when she cast off; actually she was heading for a battle with some of the Union Navy's finest battleships. Perhaps the fact that Hampton Roads was as calm as the proverbial millpond, although there was enough of a breeze to blow the smoke of the *Virginia*'s guns away, led her commanding officer Captain Franklin Buchanan, to decide to do more than make a trial run. (When cannon and small arms burned black powder the term "smoke of battle" had a literal meaning.)

For some time past the officers and seamen in the

Union ships blockading Norfolk had been hearing rumors about the rebuilt *Merrimack.* These stories had roused much curiosity, but no apprehension. On the day the *Virginia* made her first sortie no signs of preparation or extra vigilance suggested that anybody connected with the Union Navy had the slightest idea there might be a vessel within 1000 miles of Hampton Roads capable of seriously threatening any Union warship there, much less of endangering all of them. Three sailing vessels and two steam frigates—the *Congress, Cumberland, St. Lawrence, Roanoke,* and *Minnesota,* the last with a broken propeller shaft—were swinging lazily at their moorings between Newport News and Old Point Comfort; small boats were trailing from the ships' booms; freshly washed clothes were drying in the rigging, and everybody was enjoying the comparative ease of a routine Saturday. The *Cumberland*'s captain was attending a court of inquiry in the *Roanoke* and Flag Officer Goldsborough was in North Carolina with the Army-Navy Roanoke Island expedition, leaving Captain John Marston of the *Roanoke* the senior officer present.

A plume of smoke seen near Sewell's Point by a lookout in the *Cumberland* was the first thing to indicate that this day might differ in any way from many others at Hampton Roads. Since blockade runners almost never tried to get out in the daytime the *Cumberland*'s tender was rather casually sent to find out what game was afoot. On discovering a floating object resembling the top story of a house with a mansard roof the tender's crew decided they were looking at the Confederate ironclad about which there had been so much speculation. After firing half a dozen shots from her 30-pound rifle without eliciting a reply the tender returned to report to the *Cumberland*'s executive officer, commanding in the captain's absence. The long roll, beaten by the *Cumberland*'s drummer, was echoed in the *Congress,* the washing came down from the rigging, the boats were hoisted onto their davits, the two ships were cleared for action, and they signalled for the others to join them—according to a prearranged plan the Union vessels were to surround the Confederate ironclad and pound her to pieces with their heavy guns.[25]

On their way to join the *Cumberland* and *Congress* the *Minnesota* (under tow), and the *Roanoke* ran aground. Ordinarily there would have been enough water to float them over the place where they struck, but for several days past the wind had been blowing from such a direction that the sea level was abnormally low. As things turned out their stranding was fortunate.

Soon after 1 P.M. the *Congress* and *Cumberland* opened fire on the *Virginia.* Both ships scored several direct hits, but they did no damage. The *Virginia* steamed stolidly past the *Congress* to attack the *Cumberland* first because Buchanan believed she carried some rifled cannon which he feared might be able to damage the *Virginia.* While passing the *Congress* the *Virginia* fired a broadside that killed most of a gun crew in that ship. During the next few minutes, until she came within easy range of the *Cumberland,* the *Virginia* remained silent. Finally she fired two shells from her forward rifle. One of them entered the *Cumberland*'s port side near the stern and killed most of the crew of the after pivot gun; the other one practically wiped out the crew of the forward pivot gun—these were the pieces Buchanan thought were rifled, actually they were smoothbores. The *Congress* and *Cumberland* replied with their broadside batteries, delivering a hail of shot and shell that would have blown any wooden ship out of the water. But the *Virginia,* holding a steady course, rammed the *Cumberland* nearly amidships, opening a hole as big as a hogshead

25. William C., and Ruth White, *Tin Can on a Shingle* (New York: E. P. Dutton & Company, 1957), p. 39.

several feet below the water line. A shell fired by the *Virginia* at the moment she struck the *Cumberland* killed or wounded every one of the 16 men of a gun crew. By now it was obvious to everybody in the *Cumberland* that she was doomed, but a demand for her surrender was refused and she fought gallantly for another half an hour. Nearly a third of her crew of 376 officers and men had been killed or wounded before she sank with her colors still flying.

While the *Viriginia* was busy with the *Cumberland* the Confederate gunboats *Patrick Henry, Teaser, Jamestown, Beaufort,* and *Raleigh* engaged the *Congress.* Warned by the *Cumberland*'s fate, the *Congress* called upon her tender to run her ashore where at least she could not be rammed. This maneuver proved of no avail. The Confederate gunboats took positions where they could rake the *Congress* at a range of 150 yards while she could reply only with two stern guns. She withstood an hour of dreadful punishment and had her captain, most of the rest of her officers, and more than a third of her seamen killed before her flag was lowered.

After the *Congress* was surrendered two Confederate gunboats were ordered to take her officers prisoners, parole the crew, and burn the ship. As the gunboats came alongside of the frigate some Union soldiers at Newport News fired on them and the *Virginia.* Buchanan, who was standing on top of the ironclad's casemate, was wounded by a minié ball. His executive officer, Lieutenant Catesby ap R. Jones, took command of the ship, recalled the gunboats, and ordered them to pour hot shot into the *Congress* until she was ablaze from stem to stern.

In the meantime the *Roanoke* had been pulled into deep water and had wisely left the scene, but the *Minnesota* remained hard aground in spite of every effort to free her. She would have been destroyed then and there if it had not been for the fact that the tide was ebbing and the *Virginia* could not close the range to less than a mile. Finding this distance too great for accurate shooting the *Virginia* steamed back to Norfolk with every reason to expect to be able to finish her work of piecemeal destruction the next day. She could easily have done so if it had not been for the fortunate arrival of the *Monitor* or if Captain Marston had not disregarded the order he was supposed to have delivered to Lieutenant John L. Worden, commanding the *Monitor,* to proceed to Washington without stopping at Hampton Roads.

Obviously Marston acted wisely in retaining the only Union vessel that might be able to cope with the *Virginia.* However, he showed a degree of initiative not common among naval officers of his day and risked a reprimand for doing so, because, as James R. Soley, Assistant Secretary of the Navy from 1890 to 1893, said,

> [A] spirit of routine had for thirty years pervaded the naval establishment. . . . The whole tendency of the navy had been to preserve traditions, and to repress individuality in the junior officers. . . . The officers in active service, grown old in the lower grades, and but little encouraged to exercise their powers of volition, had come to regard themselves as part of a machine, and to wait for the orders of their superior[s]. As a general thing, the assumption of responsibility was neither desired nor permitted; and the subordinate who presumed, even in an emergency, to act upon his own judgment, was apt to bring down upon himself official censure.[26]

News that the *Virginia* had destroyed two of the Union Navy's finest, most heavily armed vessels without the slightest effective opposition reached Washington at 3 A.M., March 9. Nobody knew what further havoc the Confederate ironclad might wreak and, according to Welles's recollection of a Cabinet meeting held in the White House at six o'clock A.M. Sunday, the President and Secretary of War Edwin M. Stanton frequently

26. Soley, *The Blockade,* p. 7.

went to a window overlooking the Potomac to see if the *Virginia* were on her way up the river to lay Washington in ruins.[27]

Welles, who thoroughly disliked Stanton, may have exaggerated in describing him as frantic, but it is true that it was a gloomy, anxious morning in Washington and that as the news spread gloom and anxiety spread throughout the country. However, even while the Cabinet was in session a dramatic turn of events occurred at Hampton Roads.

At 7 A.M. the *Virginia,* coming down from Norfolk, opened fire on the stranded and presumably helpless *Minnesota.* A moment later the *Monitor* came from behind the frigate and steamed boldly toward the *Virginia.* The latter immediately fired her deadly bow rifle—and completely missed the *Monitor,* which was not an easy target to hit. The *Monitor* held her fire, disdainfully it seemed to eyewitnesses, until she was close aboard the *Virginia,* then both vessels fired almost simultaneously.

Both captains soon learned that their vessels' armor made it unlikely that either of them could be substantially damaged by gunfire. They also found that each craft had some advantages over the other and that each had some drawbacks. At best the *Monitor* could fire her two-gun broadside only once every seven or eight minutes. The *Virginia,* with her 10 guns, could fire at a far higher rate, but she was so slow and cumbersome that she sometimes found it impossible to bring a single piece to bear upon her nimble antagonist.

Although the *Monitor* was never intended to be used as a ram, Worden tried to disable the *Virginia* by running into her propeller. The collision brought the two ships side by side and as they lay momentarily in this position a couple of shots from the *Monitor*'s guns almost penetrated the *Virginia*'s casemate. Another hit at the same

27. Gideon Welles, *Diary of Gideon Welles* (Boston: Houghton, Mifflin and Company, 1911), 1: 62–65 passim.

spot probably would have damaged the *Virginia* badly. However, circumstances beyond their control made it impossible for the *Monitor*'s gunners to hit the same place more than once except by luck. Their only view of the world outside of the turret was through the two gunports, each just big enough to permit a gun's muzzle to protrude from it. After the guns were fired they were reloaded with the gunports closed. By the time the guns were ready to be fired again the relative positions of the two ships would, of course, have changed, making it necessary to turn the turret to bring the guns to bear upon their target. With the turret constantly moving, now in one way, then in another, the gunners soon lost all sense of direction. Because this possibility had been foreseen white marks had been painted on the deck outside of the turret to indicate the ship's sides and ends. However, these marks were soon covered with a thick layer of soot and answers from the pilothouse to the turret's frequently asked question: "Where does the enemy bear?" meant nothing to men who had no idea whether they were facing forward or aft, to port or to starboard. In these awkward circumstances the guns were run out as soon as they were reloaded, the turret was set in motion, and the executive officer, who was acting as gun captain, watched until he saw the *Virginia* at a place where he thought he could shoot at her without hitting the *Monitor*'s pilothouse, then he fired the two guns as quickly as he could, hoping they would be discharged before they swung past the target. He was, in effect, doing something like snapshooting at moving game, but with pieces weighing several tons instead of the pounds or ounces a shoulder or hand gun would have weighed. To make his task more difficult the steam engine used to turn the turret was slow both in starting and stopping.

(Ericsson claimed in 1885 that the trouble with the turret "was caused by inattention during the [*Monitor*'s] passage from New York, the working-gear having been permitted to rust from want of cleaning and oiling while exposed to the action of salt water."[28] If this were indeed the cause of the trouble it was scarcely remarkable that the *Monitor*'s crew had neglected to clean and oil the turret's machinery; they had enough to do merely to keep the ship from sinking.)

When the *Virginia* tried to ram the *Monitor* the latter's side armor protected her so well that she was simply shoved sideways as if a rowboat were pushing a plank. At the moment of collision an 11-inch solid shot fired by the *Monitor* did the most damage to the *Virginia* that she suffered throughout the battle; it struck the casemate hard enough to break its wooden backing, but not hard enough to penetrate.

Several hours after the battle began the explosion of a shell against the *Monitor*'s pilothouse temporarily blinded Lieutenant Worden. In the belief that his vessel had been badly damaged he ordered her steered into water too shallow for the *Virginia* to follow her. Before the executive officer could make his way from the turret to the pilothouse and bring the *Monitor* into action again the *Virginia* retired to Norfolk to procure ammunition and be repaired.

Each side claimed to have won a victory in the duel; actually they fought to a draw. Things might have turned out differently if the *Monitor* had used heavier charges of gunpowder, as she did later, or if the *Virginia* had used solid shot instead of shells—but she had come out expecting to have to contend only with wooden ships. However, the two vessels never met again because the Union Navy did not care to risk the loss of its only ironclad and the Confederate Navy was equally averse to chance losing the only one it had along the Atlantic coast, so both of them were ordered to act defensively, not aggressively.

---

28. Ericsson, "The Monitors," p. 291.

Early in May 1862, when Union troops from Fortress Monroe attacked Norfolk, Captain Tatnall, now commanding the *Virginia,* was told to make the best disposition he could of her. He used her first in an attempt to defend the city. When Norfolk was taken he tried to sail the ship up the James River to assist in the defense of Richmond against General McClellan's Army of the Potomac, which was fighting its way up the peninsula between the James and York Rivers. Even after her guns were removed to lighten her as much as possible she still drew too much water to get into the river so she was run ashore near Craney Island, set on fire, and abandoned during the night of May 10-11.

Many southerners blamed Tatnall for the *Virginia*'s inglorious end. A court of inquiry he requested in the hope of clearing his name decided that the ship's destruction had not been necessary at the time and place it occurred. Later a court martial, also held at Tatnall's request, acquitted him of blame and declared that he had acted wisely in the circumstances.

Two days after the *Virginia* was destroyed the *Monitor, Galena, Naugatuck* (a partly armored experimental vessel), *Aroostook,* and *Port Royal* (both wooden gunboats) started up the James to cooperate with the Army of the Potomac. They met with no trouble or serious opposition until they reached Drewry's (or Drury's) Bluff, seven or eight miles below Richmond. Finding the 200-foot-high bluff strongly fortified and the river thoroughly obstructed they anchored for the night. Early in the morning the *Galena,* followed closely by the *Monitor* and *Naugatuck,* attacked the Confederate works, hoping quickly to shell them into submission, then to proceed to Richmond.

Much was expected of the *Galena* and her captain was determined to test her fully in her first battle. He brought her smartly to anchor within 600 yards of the bluff, swung her broadside across the stream, and opened fire. A short time later she withdrew, badly damaged by the 28 hits she had sustained.

Soon after the *Naugatuck* went into action her only gun, a 100-pounder, burst. One piece of it, extending from the trunnions to the muzzle, was thrown forward onto the deck. Half of the part from the trunnions to the vent flew over the port rail. The third part, weighing about 1500 pounds, tore the coat off an officer standing nearby, but did not hurt him at all as it hurtled into the river.

A bombardment lasting more than three hours had no serious effect on the Confederate works, but did use up most of the ship's ammunition, so they withdrew, planning to return to the attack when the Army caught up with them.

If a few thousand troops had attacked Drewry's Bluff at this time they could have taken the Confederate works with the aid of the ships and the road to Richmond would have been wide open. However, at this moment McClellan gave up his peninsula campaign so the Navy's casualties and damages were incurred for no good purpose. But if the Navy had not been there to guard McClellan's line of communication until his retreat into northern Virginia was complete the Army of the Potomac probably would have been annihilated. Had this occurred the war might have ended then and there in a negotiated peace at best, in a Union defeat at worst.

After her adventure at Drewry's Bluff the *Monitor* remained idle at Norfolk until she was assigned to Captain Du Pont's squadron, which was expected soon to attack Charleston.

On Monday, December 29, 1862, the *Monitor* started for Beaufort, about as far from Hampton Roads as

Hampton Roads is from New York, in tow of the powerful side-wheeler *Rhode Island.* This voyage, like the *Monitor*'s first one, started under ideal weather conditions which soon deteriorated. By early Tuesday morning the sea was unpleasantly, but not dangerously, rough. Water often broke against the pilothouse and turret, but with the pumps easily able to take care of all that made its way below the crew felt no fear even if they were not enjoying much comfort. About 8 P.M. Tuesday, the wind suddenly increased to force 6, Beaufort scale (22 to 27 knots). Soon the *Monitor* was bucking seas 15 feet high and pitching wildly enough to bury her pilothouse under every wave. If her steering wheel had not been temporarily placed on top of the turret everyone in the pilothouse probably would have been drowned. By 10 P.M. the forward overhang had been loosened by the pounding it had taken to such an extent that the vessel was leaking too fast for pumps with a capacity of more than 3000 gallons a minute to keep the water from gaining on them. An hour later, with almost enough water in the stokehold to put out the fires, a volunteer succeeded in cutting the towline after several other men had been swept overboard and the *Monitor* steamed under the *Rhode Island*'s lee with a distress signal flying. Two of the *Rhode Island*'s boats, manned by volunteers, were sent to the rescue of the *Monitor*'s crew. When the boats reached the *Monitor* on their second trip everybody still on board was ordered to abandon ship. Several men refused to leave the turret, apparently for fear of being washed overboard, as a number of their comrades had been, and a man literally too seasick to care if the ship sank, stayed in his berth. Thus two ensigns, two assistant engineers, and 12 enlisted men were in the *Monitor* when she went down 15 miles southeast of Cape Hatteras.

5

Beginning of the River War

The steamboats *Conestoga, Lexington,* and *A. O. Tyler,* converted into gunboats by Commander Rodgers, reached the Western Flotilla's base in Cairo, Illinois, at the confluence of the Ohio and Mississippi Rivers, on August 12, 1861. At first they did little more than patrol the Ohio, the lower reaches of the Ohio's tributaries, the Cumberland and Tennessee Rivers, and the Mississippi for a short distance below Cairo, seeking to capture or destroy every river craft, not excepting rowboats, that might in any way be useful to the Confederates.

Early in September the Flotilla, now commanded by Captain Andrew H. Foote, fought its first "battle" in cooperation with some troops who were scouting along the Missouri side of the Mississippi River. As the *Tyler* and *Lexington* neared Hickman, Kentucky, they exchanged a few shots with a Confederate gunboat and a battery of field artillery. While passing Columbus, Kentucky, on the way back to their base the Union gunboats replied with a few rounds from their cannon to musket fire from the shore. A few days later the *Lexington* and *Conestoga* fought a large body of Confederate troops on the western bank of the Mississippi near Lucas Bend in Missouri. During the morning Confederate cavalrymen and field artillerymen and the gunboats followed each other down the river and back up again. In the afternoon a Confederate gunboat, oddly named the *Yankee,* took a hand in the game. Before nightfall a well-aimed 8-inch shell fired by the *Lexington* hit one of the *Yankee*'s paddle-wheels and disabled her. As the Union boats made their way back to Cairo they suffered their only casualty; one of the *Conestoga*'s men was wounded by a musket ball fired from the shore.

During the first week in October the *Tyler* and *Lexington* exchanged a few shots with some batteries at Iron Bluffs, a section of Columbus. In the next three weeks the *Conestoga* dispersed a Confederate force and silenced a battery at Canton, Kentucky, carried some troops to Eddyville, Kentucky, for an attack on a body of cavalry at Saratoga, and, accompanied by a transport, went a short way up the Cumberland River to break up a Confederate camp.

Early in November, General Ulysses S. Grant, commanding the Army of the Tennessee, sent some troops down the west bank of the Mississippi to a point across the river from Columbus to drive the Confederates out of that part of Missouri if possible, and at least to prevent reinforcements from being sent into that state from Kentucky. The following morning he received a report that his men were in danger of being captured by a Confederate force sent over from Columbus. At 6 A.M. he left Cairo with reinforcements in some transports escorted by the *Tyler* and *Lexington.* He planned to land his 3500 men on the Missouri shore out of reach of the big guns at Columbus and attack a Confederate camp at Belmont, Missouri, opposite Columbus. Once the camp was cleaned up he intended to reembark the men and return to Cairo.

The troops went ashore shortly before 8 A.M. without meeting any opposition and moved toward the camp, which contained about 2500 men. As soon as Grant's force landed the gunboats steamed downstream to bombard the works at Columbus.

After driving the Confederates to the bank of the river Grant's as yet ill-disciplined troops began looting the camp. While they were thus distracted the Confederates threw some men across the river; they rallied those who had been chased from the camp, and an attempt was made to prevent Grant's force from reaching its transports. However, the gunboats were recalled in time to hold off the Confederates until the Union troops got back on board their boats.

These skirmishes and others like them were of no great military importance, but they did accustom the crews of the gunboats and transports to the presence of enemy forces and to being under fire—useful lessons in view of the sterner work they were soon to have to do.

To defend themselves against the Western Flotilla and Grant's army the Confederates fortified Columbus so strongly that it became known as the Gibraltar of the West and built Fort Henry and Fort Donelson about 70 miles up the Tennessee and Cumberland Rivers from their confluences with the Ohio River.

Almost as soon as he took command of the Western Flotilla, Foote became convinced that Fort Henry could be attacked without too much risk and overcome without a great deal of difficulty by a joint force of gunboats and troops. Grant agreed with him to a large extent, but thought it would be better to move against Fort Donelson first. However, Grant yielded to Foote's judgment in the matter.

When the ironclad gunboats were commissioned in mid-January 1862, Foote and Grant wanted to use them to attack Fort Henry. Because orders for, and approval of, operations conducted by the Western Flotilla had to come from Army headquarters Grant proposed the movement he and Foote envisioned to General Henry W. Halleck, commanding officer of the Department of the West. Halleck called the plan preposterous and rejected it out of hand.

In view of subsequent events Halleck's cavalier treatment of Grant's suggestion seems almost amazing. However, at this time Grant was not the man as he is known now. He was, instead, a former Army captain who had served without distinction in the Mexican War, had resigned under a cloud in 1854, and had done nothing during the Civil War to attract the favorable notice of his superiors.

On January 28, 1862, perhaps at Foote's instigation, Grant again applied to Halleck for permission to move against Fort Henry and Foote wrote simultaneously to Halleck: "General Grant and myself are of the opinion that Fort Henry . . . can be carried with four ironclad gunboats and troops, and be permanently occupied. Have we your authority to move for that purpose?"[1]

Apparently Foote's backing of Grant's plan led Halleck to reconsider its merits. At any rate, Halleck telegraphed to Grant and Foote on January 29 that he would consider their proposal and decide about it as soon as he received a report concerning the condition of a road between Smithland, Kentucky, and Fort Henry. Foote, well aware of Halleck's tendency to procrastinate, replied: "I have just received your telegram in relation to Fort Henry. . . . As the Tennessee will soon fall, the movement up that river is desirable early next week (Monday), or, in fact, as soon as possible."[2]

1. *Civil War Naval Chronology* (Washington, D. C.: Naval History Division, Office of the Chief of Naval Operations, Navy Department, n.d.), pt. 2, p. 16. (Hereafter cited as *NC.*)
2. Ibid.

This message induced Halleck to tell Foote and Grant to go ahead with their plan.

Within a couple of days they assembled a fleet of transports large enough to carry half of Grant's troops to the vicinity of Fort Henry and Foote prepared the ironclads *Carondolet, St. Louis, Cincinnati,* and *Essex* and the wooden gunboats *Conestoga, Tyler,* and *Lexington* for the expedition.

Despite Foote's avowed fear that low water in the Tennessee might stymie the movement the river was in flood when the gunboats reached Panther Island six miles below Fort Henry on February 4. The speed of the current was so great that it was necessary for the heavier boats to drop two anchors apiece and keep their engines running at full speed ahead to keep them from being swept downstream and their crews were kept busy fending off huge logs brought along by the torrent. Although the men were wearied by this work, they were delighted to find that the driftwood had largely cleared the river of torpedoes.

Foote had some of these "infernal machines" fished out so they could be examined. While he and Grant were watching the *Cincinnati*'s armorer dismantle one of them on that boat's fantail it suddenly started to make a loud hissing noise. Everybody immediately left the deck for a safer place and Grant beat Foote, who was 16 years his senior, to the top of the nearest ladder. When Foote asked Grant why he had been in such a hurry the General wryly said, "The Army did not believe in letting the Navy get ahead of it."[3]

Foote and Grant made their final plans late in the afternoon of February 5. The troops were to land on both sides of the river below Panther Island and move simultaneously against Fort Henry on the east bank and Fort Heiman, a minor work, on the west side. The gun-

3. Ibid., pt. 2, p. 17.

boats would bombard the fort as soon as the soldiers were ready to go ashore.

At first Foote thought the gunboats would have to approach the fort by way of the main channel, to the east of the island, and come under fire at the extreme range of the fort's artillery—about three miles. However he discovered that the height of the river would permit the boats to pass through a chute to the west of the island, where they would be screened by dense woods until they were within a mile and a quarter of the fort. He happily took advantage of this bit of good luck.

As the gunboats emerged from the chute in line ahead about noon February 6 they took their battle positions—line abreast, with the *Essex* on the extreme right, the *Cincinnati, Carondolet,* and *St. Louis* to the left and the three wooden boats in another line abreast about a mile behind the ironclads, ready to fire over them.

The fort they were to attack was an earthwork armed with a 10-inch columbiad, a 6-inch rifle, two 42-pound, eight 32-pound, five 18-pound, and four 12-pound smoothbores. It was situated at the upstream end of a straight stretch of water nearly two miles long and 600 yards wide. During the engagement the boats steamed ahead, let the current carry them back, then steamed ahead again, thus continually varying the range while gradually closing it (on the part of the ironclads) to 500 yards.

Before the battle began Foote warned the gunners that every shot or shell they fired would cost the government $8 and he wanted none to be wasted.

> Let it . . . be distinctly impressed upon the mind of every man firing a gun [he said] that, while the first shot may be either of too much elevation or too little, there is no excuse for a second wild fire, as the first will indicate the inaccuracy of the aim of the gun, which must be elevated or depressed, or trained, as circumstances require. Let it be reiterated that ran-

dom firing is not only a mere waste of ammunition, but, what is far worse, it encourages the enemy when he sees shot and shell falling harmlessly about and beyond him.[4]

To the grim amusement of the other boats' gunners the flagship's first three shots completely missed the fort.

However, these were almost the only misses made by either side. A shell fired by a gunboat hit one of the fort's 32-pounders, wounded all of its crew, and put the piece out of action. A shot from the fort penetrated the *Essex*'s armor, killed or wounded 32 men, and disabled the boat. The *St. Louis* was hit seven times and the *Carondolet* 30 times, but nobody was hurt in either boat. The *Cincinnati* had her smokestack, her main cabin, and her small boats shot to pieces, two of her guns disabled, one man killed, and several wounded by the 32 hits she sustained.

Having the advantage of firing over known ranges the Confederates did more accurate shooting than the gunboats, but the fort was bedevilled by the worst sort of luck. The 6-inch rifle burst, killing three and wounding several more of its crew. A priming wire jammed in the vent of the columbiad and put it out of action. An accidental discharge of a 42-pounder while it was being loaded killed three men. In these circumstances the fort was surrendered to Foote before Grant's troops, delayed by muddy roads, had been fully deployed. Most of the garrison escaped capture and marched to Fort Donelson, 10 miles distant by land, nearly 200 miles by water.

The wooden gunboats, which had remained far behind the ironclads, were not damaged at all and they were sent to scour the Tennessee River immediately after the battle ended. Shortly before nightfall they reached a railroad drawbridge with its machinery sabotaged and its span closed to block the river. An hour's work by the

4. Ibid., pt. 2, p. 15.

gunboats' crews enabled them to open the bridge. Then the *Tyler,* the slowest boat, stood by long enough for some of her men to destroy the railroad tracks for a considerable distance on both sides of the river while the *Lexington* and *Conestoga* chased several Confederate steamers. When these craft were overhauled five hours later their crews set them on fire and abandoned them. The Union captains, knowing that one boat was freighted with torpedoes, another with gunpowder, stayed 1000 yards away from them. This precaution proved wise. The two boats blew up with such violence that wreckage fell back over an area half a mile in diameter and the *Conestoga*'s skylights were broken by the blast.

At this point the *Tyler* rejoined the *Conestoga* and *Lexington* and they continued steaming up the river, reaching Cerro Gordo, Tennessee, late February 7. Here they found the *Eastport,* a 280-foot-long river steamer, which was being converted into an ironclad. An attempt had been made to scuttle her, but quick work on the part of the Union sailors kept her afloat.

Leaving the *Tyler* to guard the *Eastport* and to load her with some timber and iron found nearby, the *Conestoga* and *Lexington* went on. At Chickasaw, Alabama, they seized the steamboats *Sallie Wood* and *Muscle,* the former empty, the latter loaded with iron, a commodity the Confederates could ill afford to lose.

On reaching Florence, Alabama, near Muscle Shoals (in those days sometimes called Mussel Shoals), the head of navigation, the gunboats came upon three more steamers whose crews set them on fire and abandoned them. However, a large part of their cargoes was saved, including some iron intended for armoring the *Eastport.*

As soon as the gunboats were loaded with all of the spoils they could carry they started downstream again. They reached Cerro Gordo during the night of February 8 and found that the *Tyler*'s crew had almost finished loading the *Eastport.* By this time the *Tyler*'s captain had heard of a Confederate camp at Savannah, Tennessee, so the *Lexington* was left at Cerro Gordo while the other gunboats went back up to Savannah. The camp had been abandoned before they got there, but a landing party secured a bag of mail containing some useful information and destroyed small arms, tents, provisions, etc., left behind by the fleeing Confederates. This done, the three gunboats started for Fort Henry with their prizes. Before they got there the *Muscle* sank.

The *Eastport* was sent to Cairo where she was armored and armed and added to the Western Flotilla.

Fort Donelson, which was invested by Grant's troops immediately after the surrender of Fort Henry, was built on a bluff rising 150 feet above the Cumberland River. Its guns, which commanded a long stretch of the river, were divided among three batteries. The first, about 20 feet above the water's edge, mounted nine 32-pounders and a 10-inch columbiad. The second, 30 feet higher, contained a 32-pound rifled columbiad and two 32-pound carronades. The third, on top of the bluff, had three or four heavy pieces in it.

Foote wanted to take enough time to repair the gunboats fully after the pounding they had taken at Fort Henry before they went into action again and he thought it would be helpful if he had the use of some of the mortar boats then under construction. (The so-called mortar boats were actually rafts which had to be towed from place to place and into position wherever they were used. Each of them carried a single mortar capable of throwing shells over the parapet of a fort, potentially to hit and damage magazines, barracks, officers' quarters, and other places largely safe from horizontal fire. They had walls taller than a man's height and thick enough to stop bullets from small arms.)

At the earnest request of Halleck and Grant, Foote reluctantly left Cairo on February 11 with the ironclads *St. Louis, Louisville, Carondolet,* and *Pittsburg* (all he could man from his available force) bound for Fort Donelson. Just before he departed he wrote to Welles: "I shall do all in my power to render the gunboats effective in the [forthcoming] fight. . . . If we could wait ten days, and I had men [enough], I would go with eight mortar boats and six armored boats and conquer."[5]

The *Carondolet,* sent ahead of the other gunboats, reached the vicinity of Fort Donelson on February 12. When nothing was seen or heard of the Army the gunboat announced her presence by firing a few shells—at the fort in order not to waste them.

Early the following morning the captain of the *Carondolet* received a message from Grant asking the gunboats (he supposed they were all present) to advance upon the fort at 10 A.M., at which time he would be ready to take advantage of any diversion of the Confederates' attention. In compliance with this request the *Carondolet* fired more than 100 shells at the fort. In return she received the compliments of every Confederate gun. Despite this concentrated fire she was hit only twice. However, one of the hits was made by a 128-pound shot which penetrated the forward end of the casemate, glanced over the boilers and a steam drum, then struck and burst a steam heater. With half a dozen men hurt by flying splinters the *Carondolet* dropped out of range only long enough to make repairs; then she fired 45 more shells at the fort before it became too dark for accurate shooting.

Late that night Foote arrived on the scene with the rest of the gunboats. During the morning of February 14 they prepared for the forthcoming battle by placing chains, bags of coal, lumber, etc., on their decks to pro-

5. Ibid., pt. 2, p. 22.

tect them as well as possible from plunging shots from the guns on top of the bluff.

The gunboats opened fire about 2 P.M. at a range of a mile and did most of their fighting at 400 yards. At this distance they severely mauled the fort's water battery, but suffered considerable damage from the guns of the upper batteries. A heavy shot struck the *Louisville* at the angle where the pilothouse met the deck, penetrated the casemate, and buried itself in a pile of hammocks placed in front of the boilers to protect them. A shell raked the casemate and exploded astern of the boat. A 10-inch solid shot entered a forward gunport, wrecked the gun, killed three and wounded four of its crew, traveled the length of the gun deck, and knocked a hole through the after end of the casemate before it burst. Another shell damaged the steering gear so badly the boat was put out of action. A hit on the *St. Louis*'s pilothouse killed the pilot, severely wounded Foote in the ankle, and damaged the steering gear enough to force this boat also to drop out of the fight. A 128-pound shot killed the *Carondolet*'s pilot and one of her own guns burst, although no one was hurt in this mishap. The *Pittsburg,* hit 40 times, retired from the battle with all of her guns run as far aft as possible to keep two big holes in her bow above water.

Despite the practical defeat of the gunboats Grant's troops took Fort Donelson after a siege of several days.

When Forts Henry and Donelson were overcome the Confederates precipitately evacuated Columbus, leaving large quantities of ordnance, ammunition, and stores behind them, and fell back about 45 miles to a line extending westward from Corinth, Mississippi, on the Tennessee River, to Island No. 10, a short distance above New Madrid, Missouri, in one of the Mississippi River's many S bends. (Beginning at the mouth of the Ohio River the islands in the Mississippi were numbered; a few of them also bore names.)

Soon after the Confederates abandoned Columbus 20,000 Union troops, commanded by General John A. Pope, started eagerly down the Mississippi and Foote prepared to cooperate with them, beginning at Island No. 10.

Foote realized that boats working downstream, which the Flotilla would now be doing, would if disabled drift into trouble, not away from it as had been the case at Forts Henry and Donelson. He also appreciated the tactical problems peculiar to Island No. 10. To pass the island his boats would have to navigate a winding channel close aboard a series of earthworks several miles long on both sides of the river, each of which commanded the one above it. The batteries on the Tennessee shore, or left bank, above the island, contained several 8-inch columbiads, ten 32-pound rifles, and seven smoothbores. Five more batteries on the island, facing the Missouri shore, or right bank, mounted three 8-inch smoothbores, nine 32-pounders, three 24-pounders, and four 8-inch howitzers. A floating battery, capable of being towed from place to place, carried nine more guns. New Madrid was also well armed and strongly fortified.

At this time the *Virginia*'s ramming of the *Cumberland* was a matter of recent history and everybody in the Union Navy was highly ram conscious. The Confederates were known to have six or eight gunboats in the Mississippi, some of which were equipped with rams. In view of the existence of these craft Foote felt, and the Secretary of War agreed with him, that the security of the upper Mississippi Valley depended on the Western Flotilla. In these circumstances Foote moved against Island No. 10 more cautiously than he had against Forts Henry and Donelson.

He left Cairo on Friday, March 14, 1862, with nine

gunboats and 10 new mortar boats. This force closed the northernmost of the Confederate batteries the following morning, but rain and fog kept it idle throughout that day. Sunday the mortar boats bombarded the Confederate works for many hours without accomplishing anything significant. Monday the gunboats tried their luck. The *Cincinnati* and *St. Louis,* with the underpowered *Benton* lashed between them, stayed near the left bank while the *Mound City, Carondolet* and *Pittsburg* kept to the right bank. Neither side did any grave injury to the other in the ensuing long-range artillery duel. A Confederate gun was dismounted, the *Benton* was hit four times, and the *Cincinnati*'s engines were damaged so badly that she had to be sent back to Cairo for repairs. The day's only casualties—two men killed, two mortally wounded, 11 slightly hurt—resulted from the bursting of a 42-pound rifle in the *St. Louis.* (These pieces were converted smoothbores. Because substantial amounts of metal were removed from their barrels in the process of rifling them, and they actually fired 70-pound projectiles, they often burst. It was said, only half jokingly, that they were about as dangerous to their own crews as they were to men in their target areas.)

These activities revealed a serious flaw in the design of the ironclads; the location of their paddle-wheels caused them to yaw so wildly when they were anchored by the stern that accurate shooting was impossible and they could not back upstream against the Mississippi's current. Thus they could not fight bows on as they were meant to do, but had to be tied up alongshore and used as floating batteries.

While the Navy was busy accomplishing nothing of any importance Pope's troops landed and made their way to the outskirts of New Madrid. They could have taken the town with little or no trouble, but with six Confederate gunboats moored just below it and with no shelter from their cannon it could not have been held. Accordingly the troops marched around New Madrid to Point Pleasant, 12 miles to the south, and occupied that place without meeting any opposition. Their presence there prevented reinforcements or supplies from reaching New Madrid so the Confederate soldiers were taken away from there in steamboats convoyed by the gunboats, a move Pope was unable to prevent.

Pope was now in a position to capture the Confederates on the Tennessee shore and eventually to starve out those on Island No. 10 if he could throw a force across the river. The Confederates, of course, appreciated the situation quite as well as Pope did, so they placed some heavy artillery where it could oppose any crossing. Because Pope wanted to achieve a spectacular victory such as Grant had won at Fort Donelson he asked Foote to send down one or two gunboats to overcome the Confederate batteries and ferry his troops over the river.

Foote informally solicited the views of his subordinates about Pope's request. The gunboats' officers had learned a good bit of discretion at Fort Donelson and all of them except Commander Henry Walke of the *Carondolet* opposed Pope's proposition because they thought it would involve far more risk than any possible result could justify. Foote was of the same opinion so he refused to accede to Pope's request.

Thus left to his own devices Pope had a regiment of engineers dig a canal deep enough to float his transports through 12 miles of swampland between the Flotilla's anchorage and New Madrid. (It was impossible to make the canal navigable by the gunboats with the equipment he had available.)[6]

After the Flotilla had spent two more weeks ineffectively bombarding the Confederates' northernmost bat-

6. Howard P. Nash, Jr., "The Story of Island No. 10," *Civil War Times* vol. 5, no. 8 (December 1966): 42 ff.

tery Foote called a formal council of war to discuss the question of how best to help Pope. This time Walke won Foote's reluctant permission to

> avail [himself] of the first fog or rainy night [to] drift [his] steamer down past the batteries on the Tennessee shore, and [those] on Island No. 10, . . . for the purpose of covering General Pope's army while he crosses . . . to the Tennessee side of the river [so] that he may move his army up to Island No. 10 and attack the [Confederates] in the rear while we attack them in the front.[7]

A volunteer party of 50 sailors from the gunboats *Benton, Cincinnati, St. Louis, Pittsburg,* and *Mound City,* with an equal number of soldiers from the Forty-second Illinois Regiment, daringly undertook to make things at least a little less difficult for the *Carondolet.* Setting out in five rowboats at 5:30 P.M., April 1, these men tried to reach Confederate battery No. 1, on the mainland three miles above the island, by way of the flooded woodlands along the Tennessee shore. Finding this route impassable, the boats dropped down to the southernmost mortar boat where they lay until 11 P.M. Then, keeping strict silence except for necessary orders passed in whispers and using their muffled oars only often enough to maintain steerageway, they slipped downstream in the shadows of the trees along the bank. They almost reached their intended landing place before they were discovered by a sentry who took to his heels after firing two shots at them. Fortune certainly smiled on brave men that night for no attack was made on the little force during the half an hour it spent spiking the seven guns of the battery and it made its way back to the Flotilla without a single casualty.

In the meantime the *Carondolet* was prepared as well as ingenuity could suggest for the ordeal before her. A

7. *NC,* pt. 2, p. 42.

heavy hawser was coiled around the pilothouse as high as the windows, anchor chains were placed where they would protect the most vulnerable parts of the machinery, cordwood was piled around the boilers, planks from a wrecked river barge were strewn upon the decks, a barge loaded with hay and coal was lashed along the port side to shield the magazine, and bales of hay were heaped against the weak after end of the casemate which would be exposed the moment she passed the first operable battery. To prevent the boat from revealing herself all of the guns were drawn inside of the casemate, the gunports were carefully sealed with tape to stop any leakage of light, and the exhaust steam, usually fed into the smokestacks to keep the soot in them from catching fire, was piped into the paddle box to eliminate the chuffing sound it made in the stacks.

Twenty-three sharpshooters from the Forty-second Illinois Regiment volunteered their services to help the *Carondolet*'s crew if she should be boarded. Cutlasses, pistols, and hand grenades were placed where they would be readily available if it became necessary to repel boarders and hoses for throwing hot water were rigged for the same purpose.

By the afternoon of April 4 the *Carondolet* was ready to try her luck at the first opportunity. As darkness fell that day thunder was heard in the distance. When the storm broke at 10 P.M. the *Carondolet* weighed anchor, swung into the current and started downstream with her pilot able to get his bearings only by occasional flashes of lightning. For a while it seemed possible that the weather might enable the boat to slip past the island unseen. But, just as she came abreast of the battery that had been spiked a few days earlier the soot in both of her stacks caught fire. Although the blaze was quickly extinguished, it burned long enough to reveal the boat's presence to the Confederates on the Tennessee shore and they promptly sent up rockets to warn the lower batteries.

As the *Carondolet* was about to pass the next battery her stacks caught fire again, indicating her position to every gun that could be brought to bear upon her. However, running at full speed, aided by a swift current, she was not an easy mark. To her good fortune many of the Confederate guns had been depressed as far as possible to keep the rain from entering their muzzles; before they could be elevated and trained the *Carondolet* was clear of most of them. Two shots hit the barge she was towing, but none touched the boat itself. She was actually in a good deal more danger of running aground than of being damaged by gunfire. Once she cleared a sandbar by only a few feet with the steering wheel hard over. About midnight she rounded to at New Madrid, then, with the soldiers cheering her to the echo, she ran ashore because an engineer misunderstood an order. She was quickly refloated by shifting some of her bow guns aft, whereupon all hands "spliced the main brace."

The *Carondolet* having shown the way, the *Pittsburg* ran past the island during the night of April 6-7. Soon afterward the two gunboats attacked the Confederate batteries where Pope proposed to cross the river. By noon the guns had been silenced and the troops were ferried to the Tennessee shore in transports sent down through the canal and the two gunboats. The Confederates in the vicinity beat a hasty retreat and those on Island No. 10 surrendered at midnight, April 7, to the gunboats that were still upstream.

Among the craft captured on this occasion was the C.S.S. *Red Rover* which had been damaged by mortar fire. She was sent to Cairo where she was repaired and converted into the U. S. Navy's first hospital ship. She received her first patient in mid-June. On October 1 the

Sisters of the Holy Cross volunteered for service in the *Red Rover* and the Navy Nurse Corps was founded.

Simultaneously with some of the events just described two other gunboats played a key part in one of the decisive battles of the Civil War.

After taking Fort Donelson, Grant marched his army to Shiloh, or, as it was also called, Pittsburg Landing, on the Tennessee River. To counter this move General Albert Sidney Johnston sent most of his troops to Corinth, about 20 miles southwest of Shiloh. At daybreak, April 6 (while Pope was crossing the Mississippi below Island No. 10), the Confederates surprised Grant's force which was camped in a semicircle with its ends resting on the Tennessee.

At 1:30 P.M., when the Confederates had driven the Union troops back to the river, the captain of the *Tyler* sent to ask permission of his Army superior to open fire. (He was so typical a naval officer of his day that he did not dare to act on his own initiative, even though he could clearly see what needed to be done.) Having been authorized to participate in the battle, the *Tyler* went into action at a moment when the Union left wing could not have held out much longer without her help.

About 4 P.M., after spending an hour silencing a Confederate battery, the *Tyler* communicated with General Grant, who told the gunboat's captain to use his own best judgment. Soon afterward the *Lexington* joined the *Tyler* and the pair shelled and silenced another battery. Later (5:30 P.M.) they moved to a place where they could cover a ravine between the contending armies. Just before dark they stopped a Confederate charge through the ravine with a withering fire of grapeshot and canister at point blank range. Thus the day ended with the Confederates checkmated and during the night enough reinforcements reached Grant to make his position secure.

If the Confederates had won this battle they might have been able to move into the Northwest (as the area including the present states of Ohio, Indiana, Illinois, Michigan, Wisconsin, Minnesota, and Iowa was then called) by way of the Mississippi and Ohio Rivers. At least the Western Flotilla could not have operated anywhere except on the Ohio and the upper part of the Mississippi.

After the surrender of Island No. 10 one wooden gunboat, seven ironclads, and 16 mortar boats, accompanied by Pope's transports, started down the Mississippi toward the next Confederate stronghold, Fort Pillow. The fleet anchored on the night of April 12, 1862, 50 miles from New Madrid, just below the Arkansas-Missouri state line. When five Confederate gunboats came in sight at 8 A.M. the next day the Flotilla went to meet them. After about 20 shots had been exchanged the Confederate craft retreated to Fort Pillow, 30 miles away, followed by the Flotilla and the transports.

Early in the morning of April 14 the Union mortar boats opened fire on the fort, which mounted 40 heavy guns at the top and near the base of a steep bluff.

Foote and Pope made plans to capture Fort Pillow in the same way they had taken Island No. 10. They intended to have the troops make their way down the westerly side of the river to a point below the fort and to have half of the gunboats run past it on a dark, stormy night. The fort could then be attacked from above and below at the same time. However this scheme had to be abandoned when Pope and most of his men were ordered to join General Halleck at Shiloh.

About this time Foote was forced to ask for a leave of absence because the wound he had suffered at Fort

Donelson had never healed. On May 10 he turned the Western Flotilla over to Commander Davis. Early the following morning the Confederate gunboats *Little Rebel, General Bragg, General Sterling Price, General Sumter, General Earl Van Dorn, General M. Jeff Thompson, General Beauregard,* and *Colonel Lovell* attacked the Flotilla in an attempt to relieve the pressure on Fort Pillow.

As the battle began the *Little Rebel* rammed the *Cincinnati* on her starboard quarter. The collision swung the pair side by side and while they were in this position the *Cincinnati* disabled the Confederate craft with a broadside fired with the muzzles of her guns touching her adversary's hull.

Soon afterward the *General Sterling Price* and *General Sumter* rammed the *Cincinnati,* but a tugboat pulled her into shallow water before she sank. Then, in quick succession, a shot fired by the *Carondolet* disabled the *General Earl Van Dorn* and a Confederate boat rammed the *Mound City,* causing her to sink near the Arkansas shore. Just as the hour-long battle ended an exchange of broadsides between the *Carondolet* and the *Colonel Lovell* left the latter badly damaged by a hit forward of her wheelhouse.

Davis was a plodder who liked to think about anything for a long time before he acted. Although he had no time to think before acting on this occasion, he acquitted himself well.

During the afternoon of June 5 Davis suddenly realized that Fort Pillow had been abandoned. After stopping long enough to secure the fort he made his way to Paddy's Hen and Chickens, a group of islands four miles above Memphis, where the Flotilla and the Ram Fleet anchored for the night.

At daybreak on the 6th the Union gunboats *Benton,*

Carondolet, Louisville, St. Louis, and *Cairo,* followed at a little distance by the rams *Queen of the West, Monarch Lancaster,* and *Switzerland,* met eight Confederate gunboats ranged in line abreast off Memphis. As the fleets closed, with their guns blazing, the *Queen of the West* and *Monarch* dashed through the line of Union gunboats. A moment later the *Queen of the West* struck the *Colonel Lovell* so hard as to cut her nearly in half. While still entangled with the *Colonel Lovell* the *Queen of the West* was rammed. The *Colonel Lovell* went down almost immediately, but the *Queen of the West* managed to reach shallow water on the Arkansas side of the river before she sank. The *Monarch,* close astern of the *Queen of the West,* caught the attention of the *General Beauregard* and *General Sterling Price.* They tried to ram the *Monarch* from both sides at once, but she evaded them. They collided almost head on and one of the *General Sterling Price*'s side-wheels was carried away. She limped toward the Arkansas shore where she was destroyed by gunfire. The *Monarch* now turned upon the *General Beauregard* and drove her ashore where she was abandoned. While this melee was in progress the Union gunboats shot the *General M. Jeff Thompson* to pieces, forced the *Little Rebel* ashore, and captured the *General Sumter* and *General Sterling Price.*

Although Memphis was the hub of a network of railroads and a center of river traffic, it had no land defenses of its own because it had been thought that Island No. 10 and Fort Pillow afforded it ample protection. Thus the victory at Memphis gave the Western Flotilla control over the Mississippi north of Vicksburg. However, Vicksburg turned out to be the hardest place in the Confederate States to capture except Richmond and Charleston.

As soon as the *Mound City* was raised and she and

the other Union gunboats and rams damaged at Memphis were repaired she was sent with the *St. Louis, Lexington, Conestoga* and some transports up the White River at the request of General Halleck with supplies and reinforcements for the Army of the Southwest, which was operating in the vicinity of Little Rock, Arkansas.

During the early morning of June 17 this expedition approached St. Charles, Arkansas, about 80 miles up the river. The troops landed and the gunboats steamed ahead firing at both shores. At the upper end of a bend a mile long they came upon some obstructions below a fortification standing about 200 yards back from the river bank and 75 feet above the water's edge. A well-aimed shot from this work penetrated the *Mound City*'s casemate just above a gunport. The ball killed three men in its flight then struck a steam drum. Nearly 80 men were instantly scalded to death, 25 were so severely burned that several of them soon died, 43 jumped overboard and most of them were drowned. Only 25 men out of a complement of more than 170 escaped entirely unhurt.

Despite this catastrophe the Confederate position was taken and the expedition went farther up the river to a place called Crooked Point, 63 miles beyond St. Charles, where it had to turn back because the water was dropping so fast the gunboats seemed likely to be stranded there if they stayed too long.

6
New Orleans

During the summer of 1861, about the time the Union War Department placed the contracts for the Western Flotilla's ironclad gunboats, a few men in Washington began to think about invading the Confederate States at some point along the Gulf coast—the soft underbelly of the Confederacy, to paraphrase Winston Churchill.

Assistant Secretary of the Navy Fox, who probably first suggested this plan and was certainly its chief proponent, thought the best objective of such an operation would be New Orleans. The Army officials with whom the matter was discussed preferred Mobile. Fox insisted that New Orleans, the Confederate States' biggest city and principal cotton port, was more important than Mobile. He also argued, on the basis of personal experience as both a naval and a merchant marine officer, that a fleet of wooden warships (there was no other kind yet available) could run past Forts St. Philip and Jackson, located on either side of the Mississippi River about 35 miles from the Gulf of Mexico and twice that far below New Orleans. They would then be able to capture the city, turn it over to an army of occupation, and push rapidly upstream to join forces with the Western Flotilla.

Most of the naval and military experts who heard of this idea called it ridiculous because Admiral Nelson had said that only fools would attack stone forts with wooden ships. Fox and those who agreed with him, had their belief in the practicability of such an operation strengthened by the Union Navy's experiences at Hatteras Inlet and Port Royal, where wooden ships had proved themselves to be more capable than Nelson thought they were.

Fox convinced Welles of the feasibility of the proposed operation. Welles discussed it with President Lincoln, who thought well of it. However, nothing was immediately done about it, chiefly because Fox lacked up-to-date information about the strength of Forts St. Philip and Jackson and about conditions along the Mississippi between them and New Orleans.

Early in November 1861, Commander (later Admiral) David D. Porter, who had been independently thinking about a movement against New Orleans, arrived in Washington on leave from duty with the Gulf Blockading Squadron. Porter, who had gleaned from spies, deserters, and escaped slaves the knowledge Fox lacked, outlined his plan for an attack on the city to his superiors without knowing they already had a similar idea under fairly serious consideration.

In mid-November a council of war attended by President Lincoln, Welles, Fox, Porter, and General McClellan (who had recently succeeded Scott as general in chief of the Union Armies) decided to mount a New Orleans expedition. However, the council modified Fox's plan in one important particular. At Porter's urging and with McClellan's hearty approval it was decided that Porter should be given command of a flotilla of 20 schooners, each of which was to carry a 13-inch mortar capable of discharging a 285-pound projectile. Porter promised to open the way to New Orleans by rendering Forts St. Philip and Jackson untenable, if they were not utterly destroyed, by a 48-hour bombardment with these huge guns.

(The foregoing account of the origin of the New Orleans expedition is based on an article by Gideon Welles published in the *Galaxy* in December, 1871. Porter, in his *Incidents and Anecdotes of the Civil War,* published seven years after Welles died, contradicted Welles's story and named himself as *the* originator of the expedition and *the* formulator of its successful strategy. As between Welles and Porter, I am prepared to accept the former's testimony on any disputed point. Porter was not only ever ready to sing his own praise but was also careless about facts. As one of his colleagues remarked after reading a few pages of *Incidents and Anecdotes of the City War,* if Porter wanted to believe anything the facts made no difference to him.)[1]

Welles insisted that the decision in favor of the New Orleans expedition should be kept a secret even from Secretary of War Simon Cameron to prevent word of it from being leaked to the newspapers. In the absence of censorship by the government and self-restraint by editors everything the newspapers learned was immediately published and Welles was sure that if Cameron knew of the thing the story would be broadcast. Cameron never heard of it while he remained Secretary of War. His successor, Edwin M. Stanton, did not hear of it until he had been in office for several weeks.

Since preparations for so large an expedition as the one contemplated could not be wholly concealed, talk about a movement against Pensacola, Florida, Mobile, or some place in Texas was inspired to hide the truth.

---

1. Gideon Welles, "Admiral Farragut and New Orleans," *Galaxy,* 12 (December 1871) : p. ?; David D. Porter, *Incidents and Anecdotes of the Civil War* (New York: D. Appleton and Company, 1885), pp. 63–66; Daniel Ammen, *The Old Navy and the New* (Philadelphia: J. B. Lippincott Company, 1891), pp. 461–62.

The fact that the Confederates were able to learn a lot about the formation of the Western Flotilla played into the Union's hands. The authorities at Richmond concluded that with so much shipbuilding being done upstream there was no danger to New Orleans from downstream.

However, General Mansfield Lovell of the Confederate Army, a former New Yorker, was not fooled either by the talk about other places than New Orleans or by the Western Flotilla's building program. In fact he began preparing to resist the New Orleans expedition before the Union high command decided to undertake it. On being put in charge of the defenses of New Orleans in September 1861, he asked for some heavy artillery for installation in Forts St. Philip and Jackson. His superiors, confident that there would be no attack from below the city refused to give him any more guns. The best he could do in these circumstances was to send to the forts a dozen 42-pounders from ordnance he already had on hand.

Afraid that even with these pieces added to their existing armament the forts could not prevent the passage of steamers unless they could be held under fire for some little time, Lovell had the river obstructed at a point about 500 yards below Fort Jackson, where it ran 130 feet deep, by means of a raft of cypress logs 40 feet long, four or five feet in diameter, fastened at three-foot intervals to a couple of 2½-inch iron cables stretching from bank to bank. These cables were secured to large trees on the western shore; on the other side, where there were no trees, they were shackled to heavy anchors buried deep in the ground. There were 25 or 30 of these anchors, weighing a ton and a half apiece, with 60 fathoms of chain on each one, intended to keep the raft from sagging downstream.

In March 1862, the highest water known up to that

time piled so much driftwood onto this raft that about a third of it was carried away. The gap thus created was closed by fastening seven or eight strongly built schooners to chains an inch thick which were in turn shackled to the ends of the cables. The schooners were held against the pull of the current by two heavy anchors apiece, their masts were unstepped and, with their cordage, allowed to drag astern to foul the propellers of any vessels that might attempt to break through the barrier.

When this formidable obstacle was discovered by the gunboat *Kennebec* the captain of the *Itasca* volunteered to try with the aid of another vessel, to break through it at any time Farragut wanted it done.

Back in Washington, Welles decided after much consideration that Captain (later Admiral) David G. Farragut was the only high-ranking officer to possess the necessary courage, audacity, tact, energy, self-reliance, and initiative to command the naval contingent of the New Orleans expedition.

Porter claimed, after both Welles and Farragut had died, to have nominated Farragut for this place and finally to have won Welles's assent.[2]

Welles did hesitate about choosing Farragut, and for understandable reasons. Farragut was one of the juniors on the list of captains in an organization hitherto devoted almost absolutely to the seniority system, he had not yet seen any active service in the Civil War, and he was a southerner (an east Tennessean) by birth. However, Porter's description of the part he played in the selection of Farragut is highly overdrawn. He was consulted because of his part in planning the expedition, but, as Farragut's naval secretary said when he heard of Porter's

2. Robert U. Johnson and Clarence Buel (eds.), *Battles and Leaders of the Civil War* (New York: The Century Co., 1884–1887), 2: 26–28. (Hereafter cited as *BL*.)

claim, the decision to appoint Farragut was made by Welles, with Fox and Porter merely assenting to it.[3]

In January 1862 Farragut was named flag officer of the newly created Western Gulf Squadron and ordered as soon as possible "to proceed up the Mississippi and reduce the defenses" of New Orleans. He was told to hold the city only until the expedition's troops, commanded by General Butler, could relieve him of the duty. Then, if the Western Flotilla had not already come down from Cairo, he was to take advantage of the panic it was expected the fall of New Orleans would create among the Confederates to push a strong force up the river, taking all Confederate positions on the way. Next he was to reduce the defenses of Mobile and turn them over to Butler's Army of the Gulf. With these tasks accomplished other operations of minor importance would commend themselves to his judgment and skill, but "they must not be allowed to interfere with the great object in view, the certain capture of New Orleans."[4]

The assumption that the fall of New Orleans would cause such a panic that every Confederate position along the rest of the Mississippi could quickly and easily be taken and the squadron would be able to turn its attention to Mobile and other places is wryly amusing in view of subsequent events.

Farragut's force consisted of the 40-gun frigate *Colorado* (equivalent to a 20-century dreadnaught); four 22- to 24-gun steam sloops of the *Hartford* class (battle cruisers); the side-wheeler *Mississippi* (for which there is no modern counterpart); three steam corvettes (light cruisers), carrying from seven to 10 guns each; 11 gunboats (destroyers) with two pivot guns apiece, tugboats, supply ships, colliers, etc.

---

3. Ibid., 2: 70.

4. *Official Records of the Union and Confederate Navies in the War of the Rebellion* (Washington, D. C.: Government Printing Office, 1894–1922), ser. 1, 18: 7–8. (Hereafter cited as *ORN*.)

His orders were based on the belief that there would be 19 feet of water over the bar at the Head of the Passes. At this depth the *Colorado,* drawing 23 feet, could enter the river after unloading her guns, ammunition, stores, etc., the vessels of the *Hartford* class could do so without a great deal of trouble and it would be easy for the others.

On finding an actual depth of 15 feet Farragut spent a couple of weeks vainly trying to get the *Colorado* into the river and it took all hands three weeks of strenuous labor to get the *Mississippi,* the *Hartford,* and her sister ships across the bar. Several of them had to be completely unloaded before they could be dragged over it with their keels scraping gouges a foot or so deep in the mud, and two men were killed and five hurt when a hawser being used to tow the *Pensacola* broke. To make things worse the weather was consistently stormy or foggy.

Early in April Farragut reconnoitered the forts in a gunboat. When the Confederates opened fire on her Farragut, "observing from a mast[head] remained as 'calm and placid as an onlooker at a mimic battle.' "[5]

Just before the middle of the month a Coast Survey party began mapping the river below the forts. After working under sniper fire and the forts' gunnery for five days this group provided Farragut with a reliable chart of the river, the barrier across it, the forts, and a water battery adjacent to Fort Jackson as they were at that moment.

Finally, on April 18, the mortar boats opened fire from behind a screen of trees a couple of miles below the forts. If Farragut had been consulted and heeded before the New Orleans expedition was organized the

5. *Civil War Naval Chronology* (Washington, D. C.: Naval History Division, Office of the Chief of Naval Operations, Navy Department, n. d.), pt. 2, p. 45.

U. S. Army Ram Switzerland. *Woodcut from* Harper's Weekly, *vol. 8, 1864. Courtesy Library of Congress.*

U.S.S. Carondolet. *Courtesy Naval History Division, Navy Department, Washington, D.C.*

Stephen R. Mallory, Secretary of the Confederate States Navy. Courtesy Naval History Division, Navy Department, Washington, D.C.

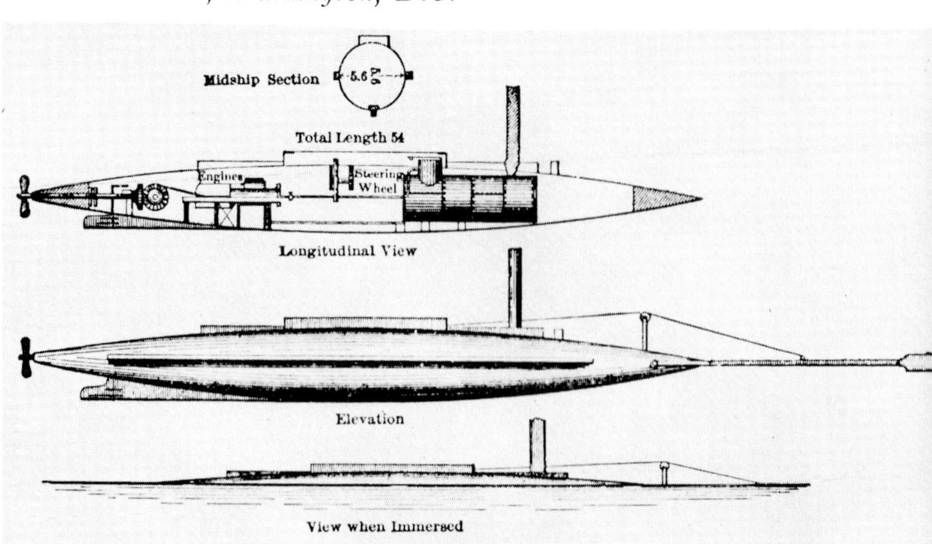

Confederate States Navy torpedo boat: David *class. From Daniel Ammen,* The Atlantic Coast. *Charles Scribner's Sons, 1883.*

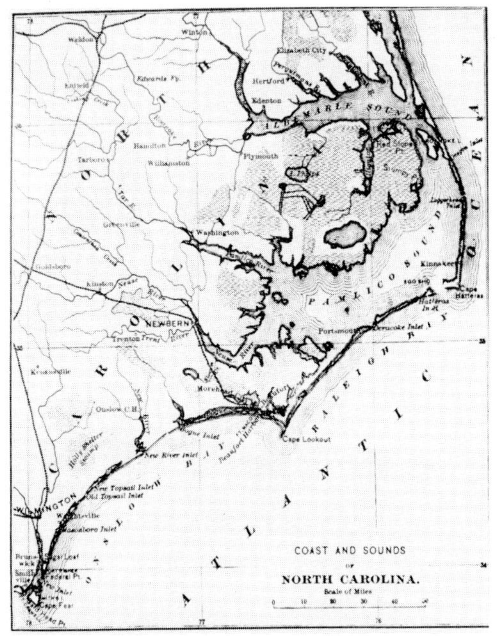

Coast and sounds of North Carolina. From Daniel Ammen, The Atlantic Coast. *Charles Scribner's Sons, 1883.*

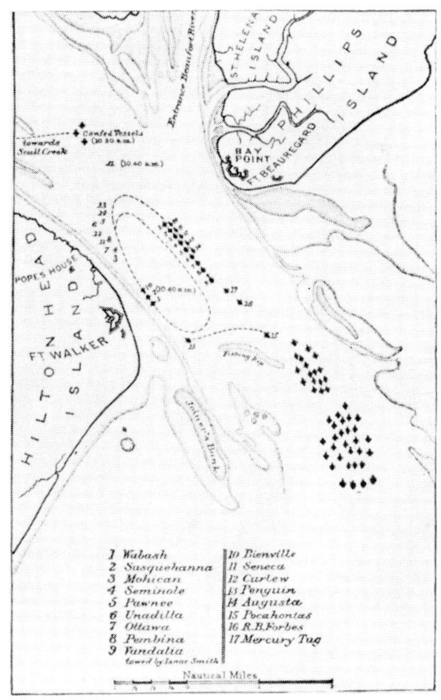

Map of the naval attack at Hilton Head, November 7, 1861. From Battles and Leaders of the Civil War. *The Century Co. 1884-87.*

C.S.S. Manassas. *Courtesy Naval History Division, Navy Department, Washington, D.C.*

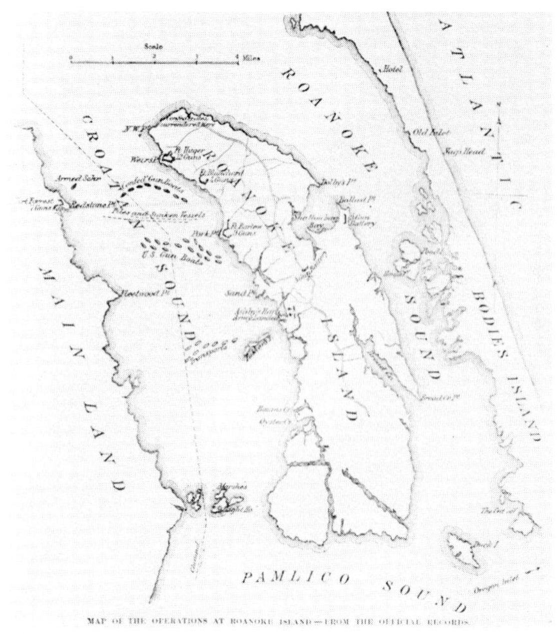

Operations at Roanoke Island. From Battles and Leaders of the Civil War. *The Century Co., 1884-87.*

Commander John M. Brooke, CSN. Courtesy Naval History Division, Navy Department, Washington, D.C.

Commander Catesby ap R. Jones, CSN, the C.S.S. Virginia's *executive officer. Courtesy Naval History Division, Navy Department, Washington, D.C.*

U.S.S. Merrimack *in drydock being converted into C.S.S.* Virginia. *From* Battles and Leaders of the Civil War. *The Century Co., 1884-87.*

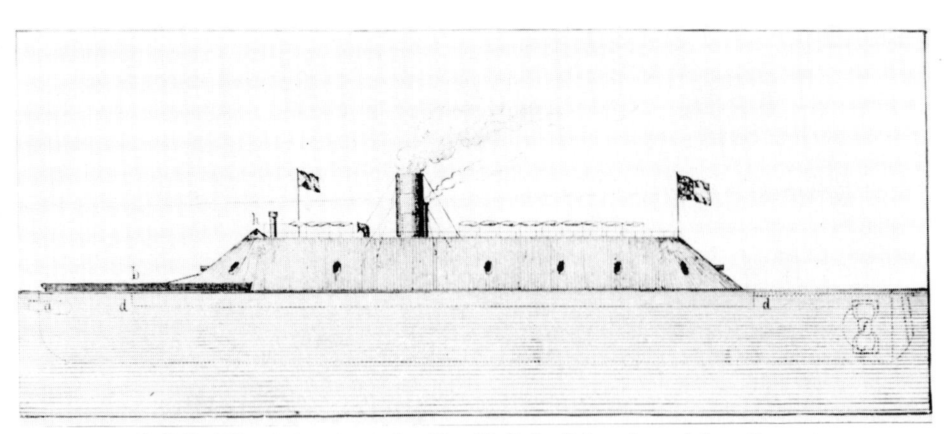

C.S.S. Virginia (*originally U.S.S.* Merrimack) *afloat in battle trim. From* Battles and Leaders of the Civil War. *The Century Co., 1884-87.*

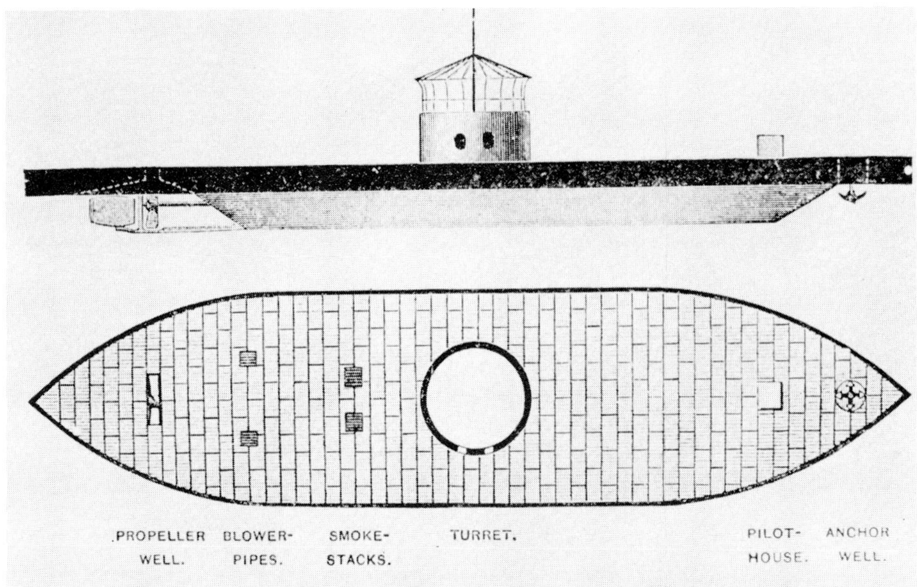

U.S.S. Monitor. *From* Battles and Leaders of the Civil War. *The Century Co., 1884-87.*

C.S.S. Virginia *sinking U.S.S.* Cumberland *at Hampton Roads, Virginia. Courtesy Naval History Division, Navy Department, Washington, D.C.*

Admiral Andrew H. Foote. Courtesy Naval History Division, Navy Department, Washington, D.C.

Union gunboats at Fort Donelson. Courtesy Naval History Division, Navy Department, Washington, D.C.

mortar boats would not have accompanied it because he did not believe they could accomplish what Porter and McClellan expected them to do. He knew (as Porter should also have known) that during the War of 1812 one poorly designed, half-finished fort, standing where there were now two carefully planned, well-built works, had turned back a British fleet after withstanding a week-long bombardment.

Two days after the mortar bombardment began (the length of time in which Porter had promised its work would be done) Farragut lost what little faith he may ever have had in its utility. At 10 P.M., April 20, he called a meeting of his ships' captains in the flagship *Hartford*. This gathering was in no sense a council of war convened to discuss possible courses of action. Farragut had made up his mind to carry out Fox's original plan of having the fleet run past the forts and the meeting was held only to permit Farragut to give instructions to those of his subordinates who were to participate in the operation.

Porter was not present because he would not be involved in the movement. However, he lodged a protest against Farragut's decision not to wait for the mortar bombardment to reduce the forts as, according to the letter of his orders, he was supposed to do. Commander James Alden read to Farragut and the others present in the *Hartford*'s cabin a lengthy paper submitted by Porter, the significant part of which said,

> Did we run the forts we should leave an enemy in our rear, and the mortar vessels would have to be left behind. We could not return to bring them up without going through a heavy and destructive fire. If the forts are run, part of the mortars [i.e. the mortar boats] should be towed along, which would render the progress of the vessels slow against the strong current. . . . If the forts are first captured, the . . . effect would be to open the way to New Orleans; whereas if we don't suc-

> ceed in taking them we shall have to fight our way up the river. . . . Nothing has been said about a combined attack by army and navy. Such a thing is not only possible, but if time permitted, should be adopted.[6]

After Porter's communication was read Farragut commented that the mortar boats could be as well protected with the fleet above the forts as with it below them; that cooperation on the part of the Army could not be effective unless some troops were landed above the forts at the same time others were landed below them; and that whatever was to be done would have to be done quickly or the ships would be reduced to a mere blockading squadron without means to make an attack since most of the shells, fuses, and material for making cartridges for the mortars had already been expended. Therefore he said,

> The forts should be run, and when a [naval] force is once above the forts to protect the troops they should be landed . . . from the Gulf side by bringing them through the bayou, and then our forces should move up the river mutually aiding each other as it can be done to advantage.[7]

(It is interesting to note that General Martin L. Smith of the Confederate Army saw the situation in exactly the same way Farragut did. Testifying before a court of inquiry concerning the loss of New Orleans, Smith said that if the fate of New Orleans had depended on the taking of Forts St. Philip and Jackson he thought the city would have been safe against an attack from the Gulf.)[8]

There may well have been a tacit understanding between Fox and Farragut about what the latter was to do if the mortar boats failed, after a fair trial, to accomplish what Porter expected of them. However, a literal

---

6. *ORN,* ser. 1, 18: 145–46.
7. Ibid., ser. 1, 18: 160.
8. Ibid., ser. 1, 6: 582.

reading of Farragut's orders could have, and one is sure would have, been used to exculpate the Navy Department if a disaster had resulted from his decision to run past the forts without waiting for them to be reduced.[9] One also feels certain that if a disaster had occurred Porter would have called upon those who heard his communication read to prove he had said "I told him so." As if to prepare the ground for just such an event Porter had already written a number of letters to his good friend Fox belittling Farragut who was Porter's foster brother as well as his superior officer.

After the ships' captains had listened to Farragut in the *Hartford*'s cabin the gunboats *Pinola* and *Itasca,* with their masts removed to make them less likely to be discovered, cleared a passage through the barrier across the river. With both vessels under a heavy fire some of the *Pinola*'s people placed a barrel of gunpowder on the third hulk away from the east bank, but the plan to fire it by means of an electric current after the ship had drifted a safe distance downstream failed because the wire broke. The *Itasca* then ran alongside of the same hulk, hooked onto it with a grapnel, and tried to tow it downstream. When the grapnel tore loose the *Itasca*'s crew fished up the chains and cut them with a cold chisel. Their sudden parting swept the gunboat down the river and against the bank. After getting free, with the *Pinola*'s help, the *Itasca* went upstream to a point well above the barrier, turned around, and steaming at full speed (to which the current added about seven knots) slammed into the middle of the chains. As she struck them her bow rose like that of an icebreaker, the chains held for a moment, then snapped and the gap was widened considerably.

Two days later, with the mortar boats almost out of

9. Cf. A. T. Mahan, *Admiral Farragut* (New York: D. Appleton and Company, 1892), p. 126.

ammunition and everybody except Porter satisfied that a lot of gunpowder had been burned to no good purpose, Farragut ordered the battleships and gunboats to prepare to run past the forts. He personally visited every vessel in the expedition to make sure his orders were fully understood. (Evidently he agreed with the Prussian military philosopher Karl von Clausewitz who said that if an order can be misunderstood it will be.) Because the current was running strongly enough to slow the ships considerably Farragut decided to wait for a shift in the wind.

The task the ships were about to face was a stern one. Fort Jackson, on the left bank of the river, shaped like a star built of stone and enclosed by a ditch, mounted three 10-inch columbiads five 8-inch guns, a 7-inch rifle, six 42-pounders, seventeen 32-pounders, and thirty-five 24-pounders. Fort St. Philip, 700 yards farther upstream on the opposite bank, made of brick and stone covered with sod, was armed with six 8-inch guns, a 7-inch rifle, six 42-pounders, a 13-inch mortar, and five 10-inch mortars. Because some of these pieces commanded the river for a distance of 3½ miles the ships (at the speed they could make against the current they would have to breast) would be under fire for 25 to 30 minutes and subject for half of that time to a raking fire first from forward, then from astern. In addition to the forts the ships would have to contend with four wooden gunboats (the *McRae,* six 32-pounders and a 9-inch shell gun; the *Governor Moore,* two rifled 32-pounders; the *Stonewall Jackson,* two 32-pounders; the *General Quitman,* two 32-pounders), and two ironclads (the ram *Manassas,* with her one gun, and the *Louisiana,* three 9-inch and four 8-inch smoothbores, two 7-inch rifles, and seven rifled 32-pounders). The *Louisiana*'s engines had not been installed so she was used only as a floating battery during the battle of New Orleans.

Everything experience or ingenuity could suggest was

done to prepare the Union ships for the forthcoming battle. They were trimmed by the head so that if any of them grounded it would not swing broadside to the current; they had lengths of anchor chain stopped up and down their sides (in lieu of armor) to protect their engines and boilers; bags of coal, ashes, and sand were placed where it was hoped they would keep raking shots from reaching the boilers and machinery; the guns' elevating screws were secured at point blank range; the ships' hulls were daubed with mud to make them less easily seen; their decks were painted white to enable the crews better to find their way around in the dark, etc.

Farragut originally intended to have the fleet advance in two columns, each to engage the fort on its own side. He decided that in the darkness and smoke vessels in parallel columns might foul each other, so he changed his plan and had what was to have been the starboard column go first.

Although Farragut was satisfied that the barrier below the forts had been cleared widely enough for the fleet to go through, some of the other officers expressed doubts that an opening really had been made in it. To settle the question the *Itasca*'s captain went in a small boat soon after dark on the night before the fleet was to move to a point well above the forts, dropped a lead line over the stern, and allowed the boat to drift past the barrier's former location. To be doubly sure he repeated this test farther from the bank before he reported to Farragut that the river definitely was clear.

Soon after 2 A.M. April 24, the flagship showed two red lights one over the other, a signal calculated not to attract any particular attention except from those specially watching for it. The fleet quickly got under way in line ahead, led by the gunboat *Cayuga,* followed in order by the *Pensacola, Mississippi, Oneida Varuna, Katahdin, Kineo, Wissahickon, Hartford* (flagship), *Brooklyn, Richmond, Sciota, Iroquois, Kennebec, Pinola, Itas-*

ca, and *Winona.* The mortar boats opened fire the moment the fleet began to move and each ship opened fire as soon as her bow guns could reach Fort Jackson. Within a few minutes smoke made it impossible to see far and nothing could be heard except the roar of heavy guns.

As the *Cayuga* made her way up the river she was brought under a raking fire from which she could do little to protect herself, came abreast of the forts, took their fire and fired into them, then suddenly found herself safely above them but under attack by three vessels—one off the starboard bow, one abeam, the third coming up from astern. A shot from an 11-inch Dahlgren, fired at a range of 30 yards, crippled the ship astern; a Parrott rifle shot took care of the one off the bow; just as the one abeam called "boarders away" the *Varuna* and *Oneida,* which had passed the *Pensacola* and *Mississippi,* came to the *Cayuga*'s help. The Confederate vessel was set on fire and driven ashore.

The *Varuna* quickly outdistanced the *Oneida* and *Cayuga* and passed the *Governor Moore,* a 900-ton ocean-going side-wheeler belonging to the Louisiana State Navy, without seeing her. The Confederate captain shot out the recognition lights at his ship's masthead and started in chase of the *Varuna.* At daybreak, with concealment no longer possible, the *Governor Moore* opened fire on the *Varuna* at a range of 100 yards. The ships raked each other several times before they came to close quarters. Then the Confederate captain, finding it impossible to bring his bow gun to bear on the *Varuna* in any other way, depressed it far enough to fire through his own vessel's deck. The shot, deflected by the *Governor Moore*'s hawsepipe, struck the *Varuna*'s funnel, but did no real damage. However, another shot fired through the hole made by the first one struck the *Varuna*'s pivot gun and killed most of its crew. A moment later the

Governor Moore rammed the *Varuna.* As the ships separated the Confederate one was hit by three 8-inch shells and set on fire. She drifted ashore with her steering gear completely disabled, her walking beam broken, 57 of her crew dead and 17 wounded.

Soon after fighting off the *Governor Moore* the *Varuna* was rammed twice by the *Stonewall Jackson.* During the 15 minutes the *Varuna* remained afloat she set the Confederate ship on fire and drove her ashore.

The *Pensacola,* following the *Cayuga,* had a comparatively easy time of it. She passed the forts without much damage and evaded an attempt by the *Manassas* to ram her.

As the *Mississippi* steamed past the forts at her best speed the *Manassas* rammed her on the port side a little forward of the mizzenmast and fired her gun. The shot did no damage; the collision caused the ship to heel about one degree, and cut a gash in her side seven feet long and a few inches deep.

The *Katahdin, Kineo,* and *Wissahickon* had difficulty in finding the opening in the barrier. The *Wissahickon* fouled one of the anchored schooners and the *Kineo* ran into the *Brooklyn,* but they all got past the forts.

As the *Hartford* closed the forts a small tug pushed a fire raft against the ship's side. In a moment the rigging was ablaze halfway up the main- and mizzenmasts, but her well-drilled crew quickly extinguished the flames. Soon afterward she ran aground close aboard Fort St. Philip, but she was able to free herself almost undamaged because, as the fort's commanding officer reported to his superiors,

> one of the columbiads had been previously dismounted, and the other could not be brought to bear; besides [the gunners'] hands were full with the other vessels coming up and the twenty-four pounder in the salient of the upper water battery,

bearing directly upon her, had been broken in two near the trunnions.[10]

By the time the *Brooklyn* came up to the forts the river was covered with a pall of smoke, punctuated by flashes of gunfire. Losing sight of the *Hartford,* on which she was steering, the *Brooklyn* ran into the barrier, fell across the stream with her bow almost touching the left bank and took a severe pounding from Fort St. Philip. Just after the ship passed the forts a steamer came down on her with the obvious intention of boarding. Veering quickly, the *Brooklyn* began firing her broadside guns as fast as they bore. Before the sternmost piece could be discharged the Confederate vessel had literally been blown out of the water. At this moment the *Manassas* rammed the *Brooklyn* almost at a right angle. As the *Brooklyn* and the *Manassas* momentarily lay beside each other a man who had appeared on the ram's deck suddenly fell overboard for no apparent reason. One of the *Brooklyn*'s quartermasters was asked later if he had seen the man vanish. "Why, yes sir," he said, "I saw him fall overboard—in fact I helped him; for I hit him alongside of the head with my hand-lead."[11]

The *Richmond,* always a slow ship, was delayed by her boilers foaming and became widely separated from the leaders. However, she engaged Fort Jackson and passed it with little loss or damage. The *Sciota* was equally fortunate, having only two men wounded.

After being badly raked by Fort St. Philip the *Iroquois* was engaged by the *McRae.* The latter's captain, Lieutenant Thomas B. Huger, who was mortally wounded at this time, had been an officer in the *Iroquois* at the outbreak of the war and had resigned to join the Confederate Navy.

---

10. F. Moore, *Rebellion Record* (New York: G. P. Putnam; D. Van Nostrand, 1861–67), 10, Docs., p. 683.
11. *BL,* 2: 67.

The *Iroquois* was the last vessel to pass the forts. The *Itasca* was disabled by a shot that hit a boiler and the *Pinola* had just reached the barrier where the *Kennebec* and *Winona* had become entangled when daybreak led Porter to recall them by signal.

Intending to return as soon as he could to cooperate with General Butler's army in reducing the forts, Farragut reassembled his ships and steamed toward New Orleans. About 10:30 A.M. at English Turn (so named because it was the place where Andrew Jackson defeated the British in the War of 1812), a few miles below the city, the fleet was delayed briefly by the Chalmette and McGee Batteries. These batteries, on opposite sides of the river, had been designed to mount more than 30 guns; they actually contained only 20 pieces because their 42-pounders had been sent, at the urgent request of the Confederate Navy Department, to arm some gunboats on Lake Pontchartrain where they were never used.

The fleet's casualties from the fire of the Forts and the Chalmette and McGee Batteries combined were 44 killed, 160 wounded; the Confederate losses were 74 killed, 72 wounded.

(After events proved the soundness of Farragut's decision to run past the forts without waiting for the mortar boats to reduce them, Porter said he had been sure all along that it could be done "with but little loss to [the] squadron.")[12]

Farragut reached New Orleans at noon April 25 and called upon the Mayor to surrender the city. The Mayor replied, in effect, that he would not surrender, but since General Lovell had stripped the place of troops it could not be defended.

Lovell's available force consisted of about 3000 militiamen, a third of them armed with muskets, the rest with

12. Ibid., 2: 38.

what he called shotguns of an indifferent sort. He knew that such a group could not offer any real resistance to the Union fleet and he marched it away to prevent the city from being bombarded. On hearing of the Mayor's gasconade Lovell offered to return with his little force, ready to fight to the last man. His heroic offer was not accepted by the Mayor.

Soon after Farragut's ships passed Forts St. Philip and Jackson, Porter called upon them to surrender to him. When this demand was politely, but firmly, refused the mortar boats bombarded the forts for several hours, then left the vicinity. According to Porter they did so to replenish their ammunition; some others suggested that they really left because of Porter's fear of the supposedly powerful *Louisiana.* (At this time nobody connected with the New Orleans expedition knew that the *Louisiana* had not been completed. However, news of the *Virginia*'s destruction of the *Cumberland* and *Congress* had been received so there was reason to fear the Confederate ironclad. If she had been in working order she probably would have made things difficult for the Union fleet.)

Throughout the bombardment by the mortars and during the battle between the ships and the forts the garrisons in both works remained cheerful, confident, and courageous. But, as the commanding officer of Fort Jackson reported:

> A reaction set in among them during the lull of the 25th, 26th, and 27th [of April], when there was no other excitement than the fatigue duty of repairing damages and . . . the rumor was current that the city [of New Orleans] had surrendered and was in the hands of the enemy.[13]

At midnight April 27, as a result of a sudden loss of

13. Moore, *Rebellion Record,* 10, Docs., 673.

morale (such as sometimes inexplicably overcomes even the bravest soldiers), Fort Jackson's garrison mutinied. Saying that the place was about to be attacked and they would be butchered if they resisted the men turned on their officers. Because it was not certain that Fort St. Philip's men would remain loyal both forts were surrendered to Porter at 2:30 A.M., April 28.

Porter's failure to arrange for the formal surrender to be made to representatives of the Army as well as of the Navy, as had been done in similar circumstances at Hatteras Inlet, aroused the ire of General Butler, a somewhat egotistical man. He angrily wrote to the Secretary of War, asking to be permitted,

> for the sake of my brave and enduring soldiers, . . . to put the truth of history right before the War Department and the country, by the simple enumeration of the fact that it was due to their efforts and that [*sic*] of their comrades, and to those alone, that Forts Jackson and St. Philip surrendered when they did. No naval vessel or one of the mortar fleet had fired a shot at the forts for three days before the surrender, and not one of the mortar fleet was within twenty-five miles at that time, they having sailed out of the river . . . for fear of the ram *Louisiana*. . . .[14]

Before the forts were surrendered Butler's troops had crossed the river and were where they could prevent the forts from receiving supplies, but his assertion that they gave in solely because of the efforts of his troops is far too strong; his statement that the mortar boats had left the river in fear of the *Louisiana* is open to doubt; but his belief that Porter's greed for praise had led him (as it often did) to overlook the accomplishments of others was well founded.[15]

14. Jesse Ames Marshall (compiler), *Private and Official Correspondence of Gen. Benjamin F. Butler During the Period of the Civil War* (Privately issued, 1917), 1: 538.
15. Howard P. Nash, Jr., *Stormy Petrel* (Rutherford, New Jersey: Fairleigh Dickinson University Press), pp. 141, 143.

Porter enthusiastically reported to the Navy Department that he had found Fort Jackson, on which the mortar fire had been concentrated, a "perfect wreck, [with] everything in the shape of a building in and about it . . . burned up by the mortar shells."[16]

In making this report Porter either displayed inexcusable professional ignorance or deliberately refrained from telling the whole truth about the condition of the forts. The mortar bombardment cost the two garrisons a total of 11 men killed and 37 wounded out of a force of 1100; nine guns out of 86 were disabled; all of the wooden buildings in both works were destroyed, but the result he had promised and which his report certainly implied had been achieved—reduction of the forts to military impotence—was not even approximated.

Among several Union Army officers who also reported on the condition of the forts were General Butler and Lieutenant (later Brigadier General) Godfrey Weitzel of the Engineer Corps. Butler said both works were "substantially as defensible as before the bombardment began, St. Philip precisely so."[17] Weitzel described Fort St. Philip as practically untouched (an important fact because it was considered the stronger of the two works) and said that while Fort Jackson might seem badly damaged to an inexperienced observer it was in fact quite as strong as it had been on the day the first shell was fired at it.

Porter never forgave Butler and Weitzel for having disagreed with his boastful report and, as we shall see, his hostility toward them affected another joint Army-Navy expedition toward the end of the war.

---

16. David D. Porter, *The Naval History of the Civil War* (New York: The Sherman Publishing Company; Hartford, Conn.: Charles P. Hatch, 1886), pp. 215, 217.
17. Marshall, *Correspondence of Gen. Butler,* 1: 428.

7

Vicksburg—1862

Farragut would have moved against Mobile as soon as possible after the capture of New Orleans if he had been free to act on his own best judgment. But, as already mentioned, his orders called for him immediately to ascend the Mississippi River. Accordingly, as soon as enough troops had reached the city to hold it, he sent part of his fleet upstream, the larger vessels under orders to go only as far as Natchez, Mississippi, the smaller ones to go to Vicksburg unless they met the Western Flotilla below there.

As the ships steamed up the river the occupants of plantations along its banks gathered to watch them pass. The white people, particularly the women, clearly showed by their looks and gestures that they would gladly have given worlds if they had possessed them to remove the "Lincoln gunboats" from the face of the earth. Negroes, specially if they were hidden from their masters' sight, unmistakably demonstrated their joy.

Baton Rouge, Louisiana, and Natchez, like New Orleans, were not formally surrendered, but there was no resistance offered at either place. However, when a demand was made for the surrender of Vicksburg the reply was: "Mississippians don't know, and refuse to learn, how to surrender to an enemy. If Commodore Farragut or Brigadier General Butler can teach them, let them come and try."[1] (Evidently the fact that Butler had been made a major general some time earlier was not yet known in Mississippi.)

By the time Farragut reached Vicksburg with the rest of his fleet in mid-May all of his vessels were in need of more or less extensive repairs. Most of them had been hit by gunfire and some of them had also been damaged by collisions that had occurred during the fighting below New Orleans or as a result of being caught in swift currents in a river with which their captains were not familiar. They were now at a place where it was extremely difficult to keep them supplied with coal and ammunition. (Besides having to buck the current for more than 400 miles supply ships had to be convoyed for most of that distance through hostile country.) The Mississippi was falling. (Already the *Hartford* had gone aground twice; once so hard she had not been freed until her coal, ammunition, and three guns had been unloaded from her.) And Vicksburg's defenses were superior to those of New Orleans.

Vicksburg, situated on a hairpin bend in the Mississippi, was defended at this time by seven 8- to 10-inch smoothbores, seven 42- and 24-pounders, and twelve rifles firing from 12- to 32-pound projectiles. These 26 pieces were mounted along a series of bluff three miles long, ranging from 150 to nearly 300 feet high. The ships could not elevate their guns enough to shoot at the Confederate batteries, even if the gun pointers could tell where they were located, but vessels coming up or down the river could be kept under a plunging fire, first from ahead, then from abeam, and finally from astern for a long time.

Nevertheless, Farragut asked General Thomas Wil-

1. *Official Records of the Union and Confederate Navies in the War of the Rebellion* (Washington, D. C.: Government Printing Office, 1894–1922), ser. 1, 18: 492. (Hereafter cited as *ORN*.)

liams, in command of the 1600 troops Butler had assigned to accompany the ships, to cooperate with him in an attack on Vicksburg. Williams, who (mistakenly, as it happens) believed the Confederates had 10,000 troops in the city and 30,000 more at Jackson, Mississippi, an hour away by train, thought such a move would be futile. Farragut could see no point in keeping his ships idling around at Vicksburg so he decided to return to New Orleans with them. He might not have done so if he had not been unwell at the moment.

On reaching New Orleans, Farragut received peremptory orders immediately to go back to Vicksburg to collaborate with the Western Flotilla in taking the city. (At this time the Western Flotilla was still above Fort Pillow, but optimism was running high in Washington.)

Farragut reached Vicksburg again with 11 battleships and 17 of Porter's mortar boats soon after the middle of June. He waited more than a week without hearing anything of or from the Western Flotilla, then decided to run past Vicksburg as he had run past Forts St. Philip and Jackson.

At 3 A.M. June 28 the ships got under way in two columns so arranged that the gunboats could fire between the cruisers, in a formation most easily visualized with the aid of a diagram:

PORT COLUMN	STARBOARD COLUMN
Itasca	
Oneida	
	Richmond
Wissahickon	
Sciota	
	Hartford
Winona	
Pinola	
	Brooklyn
Kennebec	
Katahdin	

By 6 A.M. all of the ships except the *Brooklyn, Kennebec,* and *Katahdin* had made their way above Vicksburg at a cost of seven men killed, 30 wounded, and considerable damage to the vessels. The Confederate casualties were 13 killed or wounded and the batteries suffered no damage at all.

On July 1 the Western Flotilla finally joined Farragut's fleet near the mouth of the Yazoo River about 10 miles above Vicksburg. Farragut commented when he first saw the river gunboats: "The iron-clads are curious looking things to us salt-water gentlemen; but no doubt they are better calculated for this river than our ships. . . . They look like great turtles."[2]

A few days later Farragut and Davis made a reconnaissance in the *Benton.* When a shot crashed through her side and killed a man standing near Farragut he said to Davis, "Everybody to his taste. I am going on deck; I feel safer outside."[3]

Soon after the fleet and the flotilla got together General Williams, who now had 3000 soldiers with him, put some of his men and several hundred fugitive slaves to work digging a canal across the neck of land opposite Vicksburg in the hope of enabling the ships, gunboats, and transports to bypass the city. Hard clay made this work so slow and difficult that the river had dropped too low to flood the canal by the time it was finished.

While Farragut's fleet and Davis's Flotilla lay above Vicksburg the Confederates completed an ironclad gunboat named the *Arkansas* at Yazoo City on the Yazoo River.

The construction of this vessel was authorized by an

---

2. Loyall Farragut, *The Life of David Glasgow Farragut* (New York: D. Appleton and Company, 1879), pp. 282–83.
3. Ibid., p. 286.

act of the Confederate Congress passed on August 24, 1861. Her keel was laid at Memphis the following October and she was supposed to have been ready for service before the end of the year. However, shortages of labor and material of the sort constantly experienced by Confederate shipbuilders slowed work on her and she was only partly finished when Memphis fell in June 1862. She was then towed down the Mississippi and up the Yazoo River to Greenwood City about 150 miles from Yazoo City to be completed there.

As a Confederate naval inspector said, the *Arkansas* was inferior to the *Virginia,* which she resembled except that she did not submerge her decks in action. Nevertheless, she was a formidable craft. Measuring 165 feet long, 35 feet in beam, and drawing 14 feet, the *Arkansas* had a casemate covered with 4½ inches of railroad iron except on her quarters and across her stern where there was only boiler plate. Her armament consisted of two 8-inch bow guns, one 9-inch Dahlgren, one 6.4-inch rifle, and one 32-pounder in each broadside, and two 6.4-inch rifles astern. Her speed was about six knots in slack water. Inevitably, most of her crew of about 200 men were greenhorns; her officers, headed by Lieutenant Isaac N. Brown, were from the Confederate Navy.

The finishing touches were put on the *Arkansas* at Yazoo City and she departed from there early June 14, 1862, because the river was falling, threatening to strand her. General Earl Van Dorn, in command of the defenses of Vicksburg, therefore urged that she be brought down to the city immediately. Before she had gone far it was found that a steam leak had dampened her gunpowder and made it useless. She was tied up long enough to dry the powder in a clearing beside the river and reached Haynes's Bluff, about 10 miles from the Mississippi, at midnight.

Earlier in the day a couple of Confederate deserters

had described the *Arkansas* to Farragut and Davis and told them she was on her way to Vicksburg at that moment. Farragut and Davis found it hard to believe the Confederates could have built a really formidable craft at such an out-of-the-way place as Yazoo City, but they decided to investigate the deserters' story.

The *Arkansas* left Haynes's Bluff at dawn the following day. A little later the Union ironclad *Carondolet,* accompanied by the wooden gunboat *Tyler* and the Army Ram *Queen of the West,* started up the Yazoo. With the *Tyler* leading the *Queen of the West* by a quarter of a mile and the *Carondolet* by more than a mile the trio suddenly met the *Arkansas.* Her appearance threw two of the Union captains into a virtual panic. The *Queen of the West* immediately turned tail and the *Tyler* started backing water as fast as she could. It soon became apparent that the *Tyler* would be overtaken so she rounded to and fired a broadside at the *Arkansas.* One shot badly damaged the *Arkansas*'s pilothouse, mortally wounded the pilot, and less seriously hurt three other men, including Lieutenant Brown who was standing outside and in front of the casemate. Another shot decapitated a man who, more curious than discreet, leaned out of a gunport to see what was happening. The *Carondolet* put up a brief fight before she too turned to run for the Mississippi with the *Arkansas* in hot pursuit. Just before the pair reached the mouth of the Yazoo the *Arkansas* tried to ram the *Carondolet.* The latter evaded the blow, but a broadside damaged her steering gear and she ran ashore. Brown, who believed the *Carondolet* had been sunk, continued after the *Tyler,* only a couple of hundred yards ahead of the *Arkansas.*

The sounds of this running battle were heard in the Union vessels in the Mississippi, but their officers complacently assumed that their own gunboats were shooting at Confederate field guns which were believed to be numerous on both banks of the Yazoo. Consequently none of the Union vessels stirred up their fires, banked because the weather was extremely hot. Thus they were unable to pursue the *Arkansas* as she passed them with her funnel so badly riddled that her steam pressure was down from its normal 100 pounds to the square inch to 30 pounds and her speed was only a couple of knots.

As the *Arkansas* moved slowly past the Union vessels she exchanged broadsides with at least a dozen of them. She suffered particularly from hits made by two shells and an 11-inch solid shot. One of the shells penetrated her port side, exploded, and killed or wounded most of the crew of a bow gun. The other shell, which burst just outside of a broadside gunport, killed the sponger and wounded most of the rest of the gun crew. The solid shot crashed through the side armor, killed two men and a powder boy and wounded three others of a gun crew, then killed eight men and wounded three more on the far side of the boat just as they were running the piece out to fire it.

The Union casualties resulting from the *Arkansas*'s sortie were 42 killed, 69 wounded; the Confederate losses were 14 killed, 15 wounded.

Because the *Arkansas*'s truly heroic exploit was the Confederates' first naval success in the West they rejoiced mightily over it.

General Van Dorn said Brown had "immortalized his single vessel, himself, and the heroes under his command, by an achievement the most brilliant ever recorded in naval annals."[4]

Lieutenant Brown was promoted to the rank of commander and the Confederate Congress gave him and his men a vote of thanks "for their signal exhibition of skill

---

4. *Civil War Naval Chronology* (Washington, D. C.: Naval History Division, Office of the Chief of Naval Operations, Navy Department, n. d.), pt. 2, p. 81. (Hereafter cited as *NC*.)

and gallantry . . . in the brilliant and successful engagement of the sloop of war *Arkansas* with the enemy's fleet."[5]

Mallory wrote: "Naval history records few deeds of greater heroism or higher professional ability than this achievement of the *Arkansas*."[6]

Farragut, who was much mortified by the Union ships' utter unreadiness to cope with the *Arkansas,* soon decided to run downstream past Vicksburg and to dispose of the Confederate vessel on the way.

He planned to start late in the afternoon of July 15 and Davis brought his gunboats to a place where they could bombard the Confederate batteries as a diversion. Because of unforeseen difficulties Farragut's ships did not get under way until after dark. Then, as Farragut said, he looked for the *Arkansas* with all the eyes in his head, but he could see nothing to mark her position except the flash of her guns so his fleet was unable to do her any harm.

A sharply worded message from the Secretary of the Navy about the *Arkansas*'s exploit led Farragut to propose to Davis several times that his ships coming upstream and the Flotilla's gunboats coming downstream should simultaneously attack the *Arkansas* at her mooring place under the guns of Vicksburg. Davis sensibly argued that the game was not worth the candle. However, Farragut was so insistent on doing something that Davis and Charles Rivers Ellet (son of the creator of the Army Ram Fleet) finally devised a plan which they believed would force the *Arkansas* either to go downstream where Farragut could meet her or come upstream to contend with the Western Flotilla.

Soon after 4 A.M., July 22, the gunboats *Essex* and the ram *Queen of the West* started for Vicksburg under

5. Ibid., pt. 2, p. 83.
6. Ibid.

cover of a heavy bombardment by three gunboats above the city and all of the ships below it. As the *Essex* closed the *Arkansas* the latter's bow mooring line was cut and one of her engines was started. This caused her head to swing and the *Essex,* instead of ramming the *Arkansas* as planned, grazed her side and ran ashore. The *Essex* spent 10 uncomfortable minutes getting free, but, amazingly, she had only one man killed and three wounded in that time while she inflicted a dozen casualties (six killed, six wounded) on the *Arkansas*'s crew. In the meantime the *Queen of the West* rammed the *Arkansas,* causing her to careen heavily, but not sinking her. Then the *Essex,* unable to stem the river's current, went down to join Farragut's vessels and the *Queen of the West* retreated upstream.

Farragut was satisfied that his ships could pass Vicksburg whenever and as often as he chose, but he did not think the Navy could take the place without the help of a force of 12,000 to 15,000 troops, more than General Butler had in his entire Army of the Gulf. Farragut, therefore, asked General Halleck for assistance. Halleck replied:

> The scattered and weakened condition of my force renders it impossible for me at the present time to detach any troops to co-operate with you. Probably I shall be able to do so as soon as I can get my troops more concentrated. This may delay the clearing of the river, but its accomplishment will be certain in a few weeks.[7]

In view of Halleck's inability, or unwillingness, to assist the Navy Farragut received permission to return to the Gulf with his ships.

After Farragut left Vicksburg, General Van Dorn began to dream about recapturing Baton Rouge and New

7. Farragut, *Life of David Farragut,* pp. 285–86.

Orleans. With these ends in view he ordered the *Arkansas* to aid a land force in an attack on Baton Rouge. The fact that the *Arkansas*'s engines had been badly damaged by the pounding she had recently taken was pointed out to Van Dorn. He answered that "he deemed [her] to be as formidable in attack or defense as when she defied a fleet of forty vessels-of-war, many of them ironclads," and overrode all objections to her immediate use.[8]

The *Arkansas* departed from Vicksburg early on August 3 under the necessity of steaming at full speed in order to reach her destination on schedule. Late that night, at a point 15 or 20 miles above Baton Rouge, her starboard engine broke down. Eight hours later she got under way again only to have the same engine stop within a matter of minutes. When repairs were completed the following morning her officers decided to engage a Union flotilla, even though the chief engineer warned them that the boat's machinery was not in a reliable condition. The truth of this statement was almost immediately demonstrated by the breakdown of the hitherto well behaved port engine. As she headed toward the shore to be tied up while repairs were made the starboard engine stopped running for the third time. Thus she could neither fight nor run away so she was moored to the bank, the crew opened the magazine, spread loaded shells around the gun deck, damaged the engines with hand grenades, and started a fire in the wardroom before they abandoned her. With the *Essex* shooting at her at long range the *Arkansas* blew up around noon.

Commander William D. Porter (David D.'s brother), captain of the *Essex,* wrote a glowing report of how he had destroyed the *Arkansas* in a fierce fight. Farragut and Davis disputed this story and an inquiry was begun. Porter was still defending himself and his report when he died about two years later.

8. J. Thomas Scharf, *History of the Confederate Navy* (New York: Rogers & Sherwood, 1887), p. 332.

On learning of the *Arkansas*'s destruction Van Dorn ordered Port Hudson, Louisiana, to be fortified. If he could not retake Baton Rouge he intended at least to retain control of the 200-mile stretch of the Mississippi between Port Hudson and Vicksburg so that foodstuffs could be shipped via the Red River from the Confederacy's western states to its eastern ones.

Instead of the few weeks' delay predicted by Halleck when Farragut asked for help at Vicksburg nearly six months passed before a Union force made a serious move against the city.

During this time the Union War and Navy Departments acted on a suggestion made by Captain Davis, who wrote to Welles in June 1862, saying, "that in order to acquire control of the tributaries of the Mississippi River, and to maintain that control during the dry season, it will be necessary to fit up immediately some boats of small draft [*sic*] for this special purpose."[9]

The boats designed to meet this need were officially called "light draughts" because they drew only 18 inches empty and three feet fully loaded. (It was facetiously said they could float on a heavy dew.) However, they were popularly known as "tinclads" because they were armored only against musketry except alongside of their boilers and engines where they were protected from field artillery fire. Most of them were armed with six or eight brass howitzers, a few of them also mounted a couple of light rifles as bow guns. They could transport 200 soldiers apiece and land them anywhere except immediately in front of heavily armed fortifications.

Several ironclads, including six specially designed monitors, were also added to the Western Flotilla at this time. The most powerful of these "second generation" gunboats were the 1000-ton, 280-foot-long, nine-foot draught *Choctaw* and *Lafayette,* designed by William

9. *NC,* pt. 2, p. 74.

Porter. These craft were originally commercial steamboats with their paddle-wheels located far aft and operating independently of each other to make them easier to handle in narrow waters.

The *Choctaw* was armored with 2½ inches of iron over 24 inches of oak except across the stern where the iron was only an inch thick. She had a massive forward casemate containing three 9-inch smoothbores and a 100-pound rifle. Just ahead of the paddle-wheels, where it could command the deck if the boat should be boarded, there was a smaller casemate with two 24-pound howitzers in it. Two 30-pound rifles were mounted in a third casemate behind the side-wheels. A conical pilothouse, with a couple of inches of iron over two feet of oak, stood atop the midship casemate.

The *Lafayette*'s casemate, which was quite rounded at the forward end, extended back to the paddle-wheels and had an inch of iron over an inch of gutta-percha or India rubber, backed by two feet of oak. This construction was expected to cause shot to rebound; actually the gutta-percha soon rotted and proved worse than useless. This craft had two 9-inch smoothbores at the bow, four similar pieces in broadside, and two 100-pound rifles astern. The bow and stern guns could also be used in broadside.

The *Tuscumbia,* 565 tons, the *Indianola,* 442 tons, and the *Chillicothe,* 303 tons, were similar in sillhouette to the *Choctaw*. The *Tuscumbia* and *Indianola* had twin propellers as well as side-wheels to increase their maneuverability.

The *Indianola*'s battery consisted of four 11-inch smoothbores firing ahead, in broadside, and astern.

The *Chillicothe*'s two 11-inch guns could fire only ahead or in broadside.

The *Tuscumbia* carried three 11-inch guns usable ahead or in broadside and two 100-pound rifles which could be fired astern or in broadside.

In the spring of 1862 Eads designed and built two river monitors—the *Osage* and *Neosho*. Propelled by enclosed stern-wheels, each of these vessels had a turret made of iron eight inches thick, containing two guns. They were expected to draw three and a half feet; when they were launched it was found they drew only three feet so another half inch of iron was added to their deck armor.

The last of the new river boats, built late in 1862, were the double turretted monitors *Winnebago, Milwaukee, Kickapoo,* and *Chickasaw*. These craft were driven by screw propellers and had turrets which extended below their decks and turned on six-inch iron ball bearings. Their 11-inch guns were mounted on steam-operated pistons to allow them to be lowered into the holds to be loaded.

On October 1, 1862, the Western Flotilla was transferred from the control of the War Department to that of the Navy Department, renamed the Mississippi Squadron, and placed under the command of Commander David D. Porter who was given the local rank of acting rear admiral to put him on a par with General Grant, with whom he was to cooperate in a campaign against Vicksburg.

To offset the Union Navy's building program as far as possible the Confederates extended the defenses of Vicksburg from Haynes's Bluff to Warrenton on the Mississippi River, a distance of nearly 20 miles. The rear of the city was protected by a series of swamps except at one place directly to its west.

Late in 1862 Grant and Porter launched a pincers movement against Vicksburg. Their plan was for Grant to march down the high ground east of the swamps behind the city with most of his Army of the Tennessee while General William T. Sherman, with 32,000 troops, and Porter's gunboats were to move up the Yazoo River where the Navy was to secure a beachead for the soldiers. Grant, who originated this plan, reasoned that if the Confederates met him in force Sherman could take Vicksburg with Porter's help; if the Confederates concentrated their strength against Sherman and Porter, Grant could capture the city.

This neat scheme did not work out quite the way Grant expected. A small Confederate force got in his rear and destroyed the supplies he had left at Holly Springs, Mississippi. Thus he was forced to beat a hasty retreat and to live off the country for a while.

In the meantime Sherman and Porter, wholly unaware of Grant's ill fortune, were trying to carry out their part of the arrangement.

On the morning of December 12, 1862, the tinclads *Signal* and *Marmora* reconnoitered the Yazoo River before a force of larger boats was sent into it. They found a number of torpedoes at a point about 20 miles upstream. The following day they set out to clear them away under the protection of the ironclads *Cairo* and *Pittsburg* and the Army Ram *Queen of the West.* While thus engaged the *Marmora* opened fire on a floating object. The captain of the *Cairo,* who supposed the tinclad was being attacked, went quickly to her assistance. After the situation had been explained the tinclad was ordered to proceed slowly with the *Cairo* following her. Before the *Cairo* had moved her own length a torpedo burst under her bow with such force that guns weighing several tons apiece were lifted clear off the deck. An effort to save the boat by tying her to some trees along the bank failed because the lines broke and she slid into water six fathoms deep where only the tops of her smokestacks showed. During the 12 minutes she did remain afloat her entire crew was saved, most of them being taken off by the *Queen of the West.*

The device which sank the *Cairo* was a demijohn filled with gunpowder fired by means of a friction primer and a trip wire running from a so-called torpedo pit on shore. At this time even the Confederates, who benefited most from the use of torpedoes, had not fully accepted them as legitimate weapons and the officer who had placed the one that sank the *Cairo,* and was watching when she went down, described "himself as feeling much as a schoolboy might whose practical joke had taken a more serious shape than he expected."[10]

Despite the loss of the *Cairo* the expedition pushed ahead until a bend in the river brought it before a Confederate fortification at Drumgould's (or Drumgoold's) Bluff. As soon as the tinclads had cleared away the torpedoes the *Benton* opened fire on the works at a range of 1000 yards. She had to fight singlehandedly because the river was too narrow for another boat to lie beside her.

With the gunboats in control of about eight miles of the river more than 30,000 troops landed on some low ground below Drumgould's Bluff during a heavy rainstorm. Three days later they moved against the Confederate works, but their attack was beaten back. Sherman considered the place too strong to be overcome, but he determined to hold the ground there and to send 10,000 men against another Confederate position at Haynes's Bluff, farther up the river where the gunboats could give him closer support. However dense fog prevented the movement from being undertaken, and continuing rain induced Sherman to give up operations in this vicinity. He and Porter returned to the Mississippi on January 2, 1863.

---

10. A. T. Mahan, *The Gulf and Inland Waters* (New York: Charles Scribner's Sons, 1883), p. 118.

8

Vicksburg—1863

While Porter and Sherman were busy in the Yazoo a Confederate gunboat based on Fort Hindman at the Post of Arkansas 40 or 50 miles up the Arkansas River captured a transport loaded with supplies for Sherman's troops. After their return to the Mississippi, Sherman suggested to Porter that while they were waiting for Grant to rejoin them they ought to attempt to take Fort Hindman to protect their rear. Porter agreed with this idea, but before the expedition he and Sherman planned to undertake was ready to depart General John A. McClernand, who outranked Sherman, appeared on the scene. McClernand approved the plans Porter and Sherman had made and was prepared to go ahead with them. However, Porter, who had taken an immediate dislike to McClernand, declined to have anything to do with the proposed movement unless Sherman went in command of the troops. McClernand agreed to this provision and Porter ordered his luxurious flagship *Black Hawk,* the gunboats *Louisville, Lexington, Cincinnati,* and *Baron de Kalb* (formerly the *St. Louis,* renamed when the Western Flotilla became the Mississippi Squadron), and the tinclads *Signal, Marmora, New Era, Romeo, Rattler,* and *Glide* to accompany the transports with about 30,000 troops in them.

These boats left the Mississippi on January 4, 1863, with the gunboats towed by the transports to save coal. To deceive the Confederates they went into the White River first, then used a cutoff to reach the Arkansas River.

On January 9 the troops landed four miles below Fort Hindman, a square work, 300 feet on a side, with three casemates, standing within 20 yards of the river bank, mounting three 8-inch and one 9-inch shell guns plus several smaller rifles and smoothbores. (Two of the 9-inch pieces and the 8-inch one commanded the river.)

While the soldiers were making their way to the rear of the fort the gunboats tried the range and the tinclad *Rattler* fired at some rifle pits outside of the works. Late in the day the *Louisville, Baron de Kalb,* and *Cincinnati* bombarded the fort at a range of 400 yards while the troops got into position to assault it the next day.

At 1:30 P.M. January 10 the Army was reported ready and the gunboats went into action again. Two and a half hours later the fort was surrendered to Porter just as it was about to be assaulted because its commanding officer, formerly of the United States Navy, preferred to hand his sword to a navy man rather than an army one.

The work of the Navy was never more thoroughly done than it was on this occasion. Every gun at which the boats fired was either dismounted or destroyed and the fort's casemates were knocked to pieces. However, the gunboats did not escape unscathed: The *Baron de Kalb* had one of her 10-inch guns disabled, a 32-pounder dismounted, and her hull considerably damaged; the other gunboats were hit a number of times, but their armor stood up well; the casualties, six men killed, 25 wounded in the *Louisville* and *Baron de Kalb,* were all caused by shots that came through gunports.

After Fort Hindman was taken the *Baron de Kalb* and *Cincinnati,* accompanied by the transports and troops, headed up the Arkansas River, and reached St. Charles, Arkansas, during the morning of January 14. Finding the town evacuated, the *Baron de Kalb* and the transports pushed 50 miles farther upstream to Duvall's Bluff where the railroad to Little Rock, Arkansas, crossed the river. Before the expedition had time to do anything more it was recalled by General Grant from what he called its "wild goose chase to the Post of Arkansas."[1]

A week later the gunboats, tinclads, transports, and troops were all back in the Mississippi near Vicksburg, which Grant regarded as his army's proper objective.

How to reach that objective puzzled Grant. According to the rule book he should have gone back to Memphis and established a base there as a point of departure for a fresh march down the easterly side of the Mississippi. He said in his memoirs that he did not want to do this because

> at this time the North had become very much discouraged. Many strong Union men believed that the war must prove a failure. The [congressional] elections of 1862 had gone against the [Republican] party. . . . Voluntary enlistments had ceased throughout the greater part of the North. . . . It was my judgment at the time that to make a backward movement as long as that from Vicksburg to Memphis, would be interpreted by many . . . as a defeat, . . . the draft would be resisted, desertions ensue and the power to punish deserters [would be] lost.[2]

This reasoning undoubtedly played some part in shaping Grant's attitude, but an equally compelling force may

---

1. *Official Records of the Union and Confederate Navies in the War of the Rebellion* (Washington, D. C.: Government Printing Office, 1894–1922), Ser. 1, 24: 130. (Hereafter cited as *ORN*.)

2. U. S. Grant, *Personal Memoirs of U. S. Grant* (New York: Charles L. Webster & Company, 1885–86), 1: 443.

have been that in his early childhood he developed a powerful inhibition against turning back when he had once started anywhere. This trait is said by one of his biographers to have "had in fact the formidable dimensions of an obsession" which "often dominated his actions" during the Civil War.[3]

While Grant was wondering what to do next Porter suggested that an attempt be made, with the help of the Mississippi Squadron, to flank Vicksburg's defenses by way of the Yazoo Valley. The Valley, a roughly oval-shaped area about 200 miles long with a maximum width of 60 miles, extends from just below Memphis to Vicksburg. It had so many rivers, creeks, and bayous criss-crossing it that it can aptly be likened to a huge lake with many islands rising above its surface.

Porter's idea was that by cutting the levee separating Moon Lake (at the northern end of the Valley) from the Mississippi it would be made possible for boats to go through the lake, the Yazoo Pass, the Coldwater River, and the Tallahatchie River into the Yazoo River behind Vicksburg. Although Grant later claimed he never had any faith in this scheme,[4] he agreed to try it the moment Porter mentioned it.

A couple of days' work, beginning on February 2, 1863, sufficed to cut the levee, but it took nearly three weeks for the mile-wide lowland between the river and the lake to become flooded deeply enough to float the heavier boats assigned to the expedition. Finally, on February 21, the ironclads *Chillicothe* and *Baron de Kalb,* the rams *Lioness* and *Fulton,* the tinclads *Rattler, Marmora, Signal, Romeo, Petrel,* and *Fort Rose,* with 500 soldiers in them, the towboat *S. Bayard,* three barges carrying 12,000, 10,000, and 5000 bushels of coal, and

3. W. E. Woodward, *Meet General Grant* (New York: Horace Liveright, Inc., 1928), pp. 30–31.
4. Grant, *Memoirs,* 1: 446.

transports enough for 6000 troops departed from the Mississippi.

The expedition experienced such difficulties from the winding courses of the various streams, from overhanging or fallen trees, from underwater stumps, and from shallow water that it took 10 days to make the 42 miles through the Yazoo Pass and the Coldwater River to the Tallahatchie. During this time the boats, constantly backing and filling, frequently had to fasten lines to trees along the banks to ease themselves around sharp bends. When they left the Coldwater River on March 6 all of them had suffered some damage. The tinclad *Romeo,* whose smokestacks had been swept away by tree branches, and the *Chillicothe,* which had hit a snag and was leaking badly where a plank had been started, were in the worst shape.

After a comparatively easy run down the Tallahatchie to the place where it joins the Yallabusha to form the Yazoo River the expedition came upon a sunken steamer lying above a hastily constructed fortification, called Fort Pemberton, made of cotton bales with earth piled over them. This work, armed with a 6.4-inch Whitworth rifle, three 30-pound Parrott rifles, and a number of smoothbores, was skillfully located where it could not be approached by land.

The *Chillicothe,* in the lead, was almost immediately hit four times, but only slightly damaged. The *Baron de Kalb* ranged alongside of the *Chillicothe* and the pair closed the fort with the ram *Lioness* standing by to tow them out of range if necessary. (The stream was so narrow that only two boats could work beside each other.)

Before the engagement had lasted 15 minutes a Confederate shell struck an 11-inch shell as it was being loaded into one of the *Chillicothe*'s guns. The resulting explosion killed two men instantly and wounded 11 others, two of them mortally. Both boats then withdrew long enough to put some cotton bales in front of their forward casemates. Soon after they went back into action the *Chillicothe* was so badly damaged it took 24 hours of hard work to repair her.

At the end of five more days, each of them more or less like the one just described, nine of the gunboats' men had been killed and 30 wounded, but the fort was quite as strong as it had ever been. Since there was obviously nothing to be gained by staying where they were the expedition's commanders decided to break off the attack and retrace their route to the Mississippi.

On their way back they met General I. F. Quinby coming up with reinforcements sent by Grant under the impression that the original force was doing well. Quinby who thought Fort Pemberton could be taken if a really determined effort were made, ordered the troops to return with him. At his request the gunboats went along too. Two weeks later Quinby retired without having accomplished anything more than the original force had.

The slow progress of the Yazoo Valley expedition led Porter to have the gunboats *Louisville, Cincinnati, Carondolet, Mound City,* and *Pittsburg,* four monitors, four towboats, and the ram *General Sterling Price* make an attempt to reach the Yazoo above Haynes's Bluff by a route about 200 miles long through Steele's Bayou, Black Bayou, Deer Creek, and the Big Sunflower River. And, as he said, he "omitted nothing that might possibly be wanted" to assure the success of this expedition;[5] he even commanded it personally.

Everything went smoothly until the boats reached Black Bayou, four miles from Deer Creek. Here it became necessary for the ironclads to force a passage for themselves and the other boats by pushing trees aside or

5. David D. Porter, *Incidents and Anecdotes of the Civil War* (New York: D. Appleton and Company, 1885), pp. 144–45.

pulling them up by their roots. Twenty-four hours of what Porter called "terrible work" got the expedition through Black Bayou to Deer Creek where no more difficulty was expected. But instead of affording easy sailing, Deer Creek proved to be narrow, tortuous, shallow, and strewn with willow trees with their roots under water. Three more days spent toilsomely hacking away interlaced tree branches and roots enabled the boats, moving at speeds of from half a mile to a mile an hour, to reach Rolling Fork on March 21. Porter already knew that the Confederates were chopping down trees ahead of the expedition; he now learned that they were doing the same thing behind it. He quickly decided to return to the Mississippi while it was still possible to do so. Since there was not enough room where they were for the gunboats to turn around their rudders were unshipped to keep them from being smashed to pieces and the boats backed upstream, bouncing from tree to tree, as Porter put it.

Just before the Yazoo Valley expedition departed Porter sent the Army ram *Queen of the West* below Vicksburg to prevent the shipment of supplies from Arkansas, Louisiana, and Texas to the eastern Confederate States by way of the Red River. He was emboldened to do this because with Union troops holding the western bank of the Mississippi for several miles south of Vicksburg the boat would have a secure base below the city. He selected the *Queen of the West* for the task because he believed she would be well able to cope with any opposition he thought she was likely to meet and, having no guns, she could not be used to fight against shore batteries above Vicksburg.

The Army ram's commanding officer, 19-year-old Colonel Charles R. Ellet, was ordered to sink the Confederate steamboat *City of Vicksburg* on his way downstream.

Experience had taught the Union captains that in any such run as the *Queen of the West* was about to undertake the pilot would be a prime target. For this reason the steering wheel was relocated in a more protected position than it usually occupied. Immediately after the boat got underway at 4:30 A.M., February 2, 1863 (the day the levee was cut to permit the Yazoo Valley expedition to enter Moon Lake) it was found that the steering wheel did not work well in its new position. She was, therefore, anchored while the wheel was returned to its original place. It took so long to make this change that the sun had risen by the time the *Queen of the West* rounded the point above Vicksburg. She was, of course, quickly discovered and brought under a heavy fire, but she was hit only three times before she came abreast of the *City of Vicksburg* which she intended to sink.

Either by luck or because the Confederates had profited by experience when the *Essex* had attacked the *Arkansas* the *City of Vicksburg* was moored in a position which made it necessary for the *Queen of the West* to round to in order to ram and at the moment of collision the current caught her stern in such a manner that, with her bow acting as a pivot, she swung rapidly around and lost most of her headway. As this happened a shell set fire to some cotton bales near her starboard paddle-wheel. The attempt to sink the *City of Vicksburg* was necessarily abandoned, the burning bales were cut loose, and the *Queen of the West* scurried downstream under a rain of shot that knocked her cabin to pieces and made a couple of holes through her hull, but caused no serious casualties.

As soon as the *Queen of the West* was repaired she steamed down the Mississippi and went a short distance up the Red River before a shortage of fuel forced her to return to her base with Grant's troops below Vicksburg.

During this time the *Queen of the West* captured the

steamboat *A. W. Baker,* which had just delivered her cargo to Port Hudson; the steamer *Moro,* loaded with 110,000 pounds of pork, 500 hogs, and a large quantity of salt, destined for Port Hudson; and the steamer *Berwick Bay,* carrying 200 barrels of molasses, 10 hogsheads of sugar, 30,000 pounds of flour, and 40 bales of cotton. She also destroyed 25,000 pounds of metal ready to be shipped to Port Hudson.

Pleased with what he had accomplished, but satisfied that he could have done much more if he had been able to cruise longer, Ellet proposed to come upstream past Vicksburg in the *De Soto,* a small, fairly fast, shallow draught steamboat recently captured by the Union troops below the city and to tow a barge loaded with coal back with him. Porter vetoed this proposal on the grounds that it would be unnecessarily risky. Instead he let an unmanned barge drift downstream to be picked up by the *Queen of the West.*

Thus furnished with fuel enough to last for a month the *Queen of the West* and the *De Soto* departed on a cruise.

During the morning of Thursday, February 12, the *Queen of the West* turned from the Red River into the Atchafalaya hunting for a transport Ellet had heard was in the vicinity. When the ram came upon a lightly guarded wagon train four miles upstream a landing party unharnessed the mules, scattered them, and destroyed the wagons. A search lasting most of the rest of the day failed to discover the transport, but the *Queen of the West* did find and shoot up another wagon train. Toward nightfall, near the place where the first wagon train had been encountered, a musket shot fired from the shore wounded one of the ram's officers. The following day a landing party burned all of the plantation buildings along the Atchafalaya from Simsport, Louisiana, to the Red River on the grounds (probably well taken) that some of their occupants had been among the guerrillas who fired on the *Queen of the West.*

When Ellet left the Atchafalaya he intended to steam down to Port Hudson in hope of communicating with Farragut below there. For some reason Ellet changed his mind and turned up the Red River instead of going down it to the Mississippi. After spending a night at anchor off the mouth of the Black River, which was too narrow for the *Queen of the West* to enter, she and the *De Soto* started for Gordon's Landing where Ellet had heard there were three steamers he could easily capture.

At 10 A.M. Saturday, about 15 miles above the Black River, the *Queen of the West* and *De Soto* encountered the *Era No. 5,* a fine steamer owned by the Red River Packet Company. Captured after a short chase, the *Era* was found to have 4500 bushels of corn, two Confederate Army officers, a few soldiers, and some citizens on board. The officers were held as prisoners of war, the soldiers and citizens were quickly put ashore, the soldiers under parole.

Leaving the *Era* with a few men to guard her and the two prisoners, the *Queen of the West* and *De Soto,* guided by the *Era*'s unwilling pilot, steamed as rapidly as the winding channel permitted toward Gordon's Landing. As the *Queen of the West* rounded a bluff near the landing the pilot deliberately ran her aground within easy range of a newly constructed fort mounting four 32-pounders. The *Queen of the West*'s engines and boilers were soon so badly damaged as to leave no hope of getting her free and her crew panicked. Some of them seized the only small boat and left the others, including Ellet, to make their escape on floating cotton bales. Because there was no way to remove a wounded officer the *Queen of the West* was not destroyed as she should have been.

Just as the last of the *Queen of the West*'s people

reached the *De Soto* a dense fog covered the river. Ordinarily the *De Soto* would have stayed where she was until the fog lifted; in the existing circumstances she started downstream as fast as she could go. She had not gone far before she ran aground and unshipped one of her rudders. This accident made her almost unmanageable. During the three hours it took her to get to where the *Era* had been left the *De Soto* sometimes faced downstream, at other times upstream, without much regard to her helmsman's efforts to steer her. Boarding the *Era,* the harried men set the *De Soto* on fire, then all hands not engaged in operating the *Era* threw her cargo of corn overboard to help her gain speed as, despite thunder, lightning, and fog she raced for the Mississippi hoping to reach Vicksburg ahead of the boats Ellet knew would soon be pursuing her.

Meanwhile, Porter had decided to double the strength of the Red River blockade by sending the ironclad *Indianola* down to join the *Queen of the West.* After being delayed 18 hours by fog the *Indianola* got under way at 10:15 P.M., February 12 (the day the *Queen of the West* entered the Atchafalaya River). She successfully passed Vicksburg with coal barges lashed on either side to assure her and the other boats below the city of an ample supply of fuel.

When Porter learned that the *Indianola* had safely passed Vicksburg he sanguinely wrote to the Secretary of the Navy:

> This gives us entire control of the Mississippi, except at Vicksburg and Port Hudson, and cuts off all supplies and troops from Texas. We now have below [Vicksburg] 2 XI-inch guns, 2 IX-inch guns, 2 30-pounder rifles, 6 12-pounders, and 3 vessels.[6]

---

6. *ORN,* Ser. 1, 24: 375.

By the time this letter reached Washington the Confederates had regained control of the Mississippi between Vicksburg and Port Hudson.

On February 16 the *Indianola,* heading down the river, met the *Era* coming upstream to report the loss of the *Queen of the West* to Porter. The *Indianola*'s captain assumed that another boat would be sent to cooperate with him as soon as Porter heard the news so he continued on his course toward the Red River. Because Porter was away with the Steele's Bayou expedition and his subordinates had found they were wise not to be responsible for anything to which he could possibly take exception no other boat was sent to join the *Indianola.*

Three days after he spoke the *Era* the *Indianola*'s captain learned that the Confederates were planning to attack his boat with the *Queen of the West,* which had been fully repaired, the ironclad gunboat *Webb,* and two other boats. To avoid a fight against such heavy odds the *Indianola* promptly started upstream. She tied up at a point about 35 miles below Vicksburg during the afternoon of February 24. Just before she cast off several boats were seen coming up the river. Although the *Indianola* could have gained considerable speed by dropping the coal barges she had with her, the captain kept them because he thought the fuel they held would be needed by the boats he still expected would be sent to join him. Thus encumbered the *Indianola* was unable to keep ahead of her pursuers. She was cleared for action and rounded to in order to take advantage of the current soon after 9:30 P.M. (The official report made by the captain does not mention the afternoon stop, but implies that she had been constantly under way before being overtaken. A newspaper correspondent who was on board was unable to say why the stop was made.)[7]

7. F. Moore, *Rebellion Record* (New York: G. P. Putnam; D. Van Nostrand, 1861–67), vol. 6, docs., 423, 425.

The *Queen of the West* opened the battle by ramming the *Indianola,* but the coal barge she had alongside saved the gunboat from any damage. Then the *Indianola,* steaming at full speed, assisted by the current, and the *Webb,* bucking the current, slammed into each other with an impact that knocked down nearly everybody in both boats, but amazingly did no great damage to either of them. As the dogfight continued the *Indianola* was rammed several more times by the *Webb* and at least once more by the *Queen of the West.* Finally, like a large man in a rough-and-tumble fight with several smaller ones acting in concert against him, the *Indianola* was overcome. Soon after 11 P.M. she was run ashore and abandoned on the west bank of the river in the hope that she could be saved by Union troops. However, the Confederates managed to tow her to the other shore where she sank in shallow water.

The Confederates were not destined to make any use of the *Indianola.* A day or two after she was captured some prankish Union soldiers, who had no knowledge of her loss, made an imitation monitor out of a coal barge with a hogshead for a turret and logs for guns. This "ironclad" was set adrift above Vicksburg and it passed the Confederate batteries defending the city despite having been hulled several times. The appearance of this "monster" so badly frightened the men who were trying to salvage the *Indianola* that they blew her up.

When Farragut learned of the loss of the *Indianola* he decided it was time for him again to take a hand in the game being played along the Mississippi. Selecting four battleships and three gunboats—poorly suited for blockade duty because they were rather slow but well adapted for the work he had in mind because of their heavy batteries—he departed from New Orleans on March 12, 1863, for his third voyage up the river. The following day these vessels anchored a few miles below Port Hudson, then besieged by the Army of the Gulf, now commanded by General Nathaniel P. Banks.

Farragut decided, after conferring with Banks, to run past Port Hudson under cover of a bombardment by all of the mortar boats in the vicinity and two gunboats. Banks promised simultaneously to attack the Confederate works with 12,000 troops, all he felt he could spare from other duties.

Port Hudson, situated on the easterly side of the Mississippi, opposite Thomas Point where the river made a hairpin turn, was defended by two 10-inch columbiads, two 8-inch columbiads, two 42-pounders, three 24-pounders, and eight rifles mounted along a series of bluffs four miles long and from 80 to 100 feet high. At the foot of these bluffs there were a number of reflecting lamps of the sort used as headlights on locomotives and large piles of wood ready to be ignited to illuminate the river. Farragut did not intend even to try to reduce the Confederate works. He wanted only to get his ships past them with the least possible damage in order to have an efficient naval force between Port Hudson and the south side of Vicksburg.

After what Farragut called "a free interchange of opinions on the subject, every commander arranged his ship in accordance with his own ideas." One of the things Farragut did was to have a speaking tube rigged from the *Hartford*'s mizzentop to her quarter-deck. Thus the pilot, stationed in the top where he could see over the smoke, and Farragut, who usually stayed in the mizzen rigging during a battle, could more easily communicate with the ship's captain and the quartermasters.[8] This device remained a part of the ship's equipment.

At this time Farragut adopted a tactic he was to use

8. Loyall Farragut, *Life of David Glasgow Farragut* (New York: D. Appleton and Company, 1879), p. 323.

again. Three of the bigger ships and the three gunboats were lashed side by side so that one could help the other if either should be disabled. The gunboats were secured on their consorts' port sides as far aft as possible to leave the larger vessels' port broadsides fairly clear for use against a Confederate battery on the right bank of the river above Thomas Point. The pairs were made up with the speeds of the battleships and gunboats in mind. Thus the *Richmond,* the slowest ship, was accompanied by the *Genessee,* the fastest, most powerful of the gunboats. The other pairs were the *Hartford-Albatross* and the *Monongahela-Kineo.* The *Mississippi* traveled alone because her side-wheels made it too awkward for her to take another vessel alongside.

When the vessels got under way at 10 P.M., March 14 the *Hartford* and *Albatross* led the way in handsome style, stopping whenever the smoke became too dense for the pilot to see far enough ahead to suit him. As the pair rounded Thomas Point the current swung them almost ashore, but with the *Albatross* backing hard while the *Hartford* maintained full speed ahead they got back on course and steamed out of range of the Confederate batteries with only two men killed, two wounded, one lost overboard, and no serious damage to either vessel. Perhaps deceived by the reports of the howitzers in the *Hartford*'s tops, which were nearly on a level with the lowest bluffs, the Confederates did not depress their guns far enough to hit the ship as often as they did those following her.

Although the *Richmond* and *Genessee,* second in line, were hampered by thick smoke, they made their way safely to Thomas Point and were about to round the turn there by midnight. Then a plunging shot from a 6-inch rifle hit the *Richmond* several feet above the berth deck, penetrated a barricade made of hammocks, clothing, etc., damaged a safety valve, passed through a smoke pipe near a steam drum, and glanced upward to smash the safety valve on another boiler. Within a few minutes the stokehold was full of smoke and the steam pressure dropped to nine pounds, bringing the engine to a stop. When it proved impossible for the *Genessee* to tow the *Richmond* against the current the pair retreated. The casualties in these two vessels were three killed and 12 wounded, including the *Richmond*'s executive officer who was mortally hurt.

As the *Monongahela* and *Kineo* neared Thomas Point a shot lodged between the gunboat's sternpost and rudderpost, making it impossible to steer her. However, the pair pushed on until the *Monongahela*'s pilot lost his bearings in the smoke and the ship ran aground at full speed. When she struck the bank the gunboat tore loose and ran ashore. Half an hour later the *Kineo* got free and, despite the fact that she could not be steered, she managed to pull the *Monongahela* into deep water.

Since the *Kineo* could be of no more help she was cast loose to drift back past the batteries while the *Monongahela* kept on her way. She had almost rounded Thomas Point when an overheated crank pin made it necessary to shut down her engine and she drifted out of action. (The pin had first given trouble when the safety valve was tied down to permit raising double the normal steam pressure and the engine was worked far beyond its rated capacity in an effort to back the ship free after she grounded. By loosening the bearing and playing a hose on it the engine was kept going for a while.)

The *Kineo* was fortunate enough to have suffered no casualties in this affair; six men were killed and 21 wounded in the *Monongahela*. Among those hurt was the captain who was standing on the bridge when it was shot away.

The *Mississippi,* last in line, came up to Thomas Point without much trouble; then, just as all on board were congratulating themselves on their good luck, the pilot lost his way in the smoke and the ship ran ashore at full speed. With the steam pressure up from its normal 13 pounds to 25 pounds—all the chief engineer thought the boilers could stand—the engine was kept in reverse for 35 minutes without moving the ship an inch. By this time she was under such a heavy fire that she had to be abandoned. Her three small boats still capable of floating were used to land the crew on the west bank of the river where the Union gunboat *Essex* picked them up. As water poured into the *Mississippi*'s stern through numerous shot holes she slid away from the bank and drifted downstream ablaze from stem to stern. About 3 A.M., just before she sank, her port guns, which had been left loaded, grew hot enough to fire a last defiant broadside.

Upon mustering the ship's company 64 men of a total of 297 were found missing; of these 25 were believed to have been killed.

Uncertain what fate had overtaken the rest of his ships Farragut tried vainly to signal to them from above Thomas Point during the morning of March 15.

Heading upstream the *Hartford* and *Albatross* engaged a battery of four rifled pieces at Grand Gulf, Mississippi, but encountered no other opposition and reached the vicinity of Vicksburg on March 20.

As soon as Farragut was able to communicate with the Union Army he sent a message to Admiral Porter asking him to send down one or two ironclads and a couple of rams to enable the *Hartford* and *Albatross* more efficiently to patrol the river. However, Porter was away with his Steele's Bayou expedition and the senior officer present refused to take the responsibility for detaching any vessels from the Mississippi Squadron.

By chance General Alfred W. Ellet of the Army Ram

Fleet learned that Farragut had applied to the Mississippi Squadron for help. When Ellet asked various naval officers about the matter none of them would answer him.

A little later Ellet received a letter from Farragut asking him to call on board the *Hartford*. When the two met Ellet asked Farragut if he would like to have some rams sent down to him. Farragut answered that he would, but he did not want to interfere with Porter, under whose command he understood the rams were. Ellet said all he wanted to know was did Farragut believe it would benefit the Union cause to have a couple of rams below Vicksburg. Farragut replied: "Certainly." Ellet promised that two would come down that night.[9]

Before their conference ended Farragut, possibly on his own initiative, probably at Ellet's request, gave the General a letter reading:

> I have written to Admiral Porter to the effect that I am most desirous of having an ironclad gunboat and two rams below Vicksburg to maintain control of the [Mississippi] River between this place and Port Hudson. . . . I am unwilling to interfere with the Admiral's command in any way, but I feel sure that if he was here he would grant the assistance I so much need to carry on this great object. I beg to assure you that nothing would be more gratifying to me than to have two of your rams.[10]

As soon as Ellet returned to his headquarters he ordered the *Lancaster* and *Switzerland* to be made ready to join Farragut's force. He unsuccessfully sought to have an ironclad sent with them.

The necessity of loading the rams with large amounts of stores delayed their departure almost until daybreak. Then, although they ran without lights, they were quickly discovered. When the Confederate batteries opened fire on them they sought safety in speed. Fifteen or 20 shots

9. Ibid., p. 349.
10. *ORN*, Ser. 1, 24: 24.

missed the *Switzerland* before a shell plunged into one of her boilers, stopping the engines immediately. However, she drifted safely past Vicksburg and was taken in tow by the *Albatross*. The *Lancaster* was less fortunate. One shot burst a steam drum; another struck the boat's stern, traveled the length of the hull, and made a big hole in the bow; a third tore the steering wheel to pieces, but amazingly left the pilot unhurt. The *Lancaster*'s crew abandoned her moments before she sank. All of those who took to her small boats were saved; one man who tried to swim ashore drowned.

When Porter returned from Steele's Bayou he berated Ellet for the loss of the *Lancaster* and said the *Switzerland* was not fit for service at any distance from a machine shop. Actually the damage she suffered was fully repaired within four days.

Porter and Ellet also corresponded acrimoniously about who commanded whom for some little time, but Farragut had the ram.

Early in April, Farragut dropped down to Port Hudson with his little fleet hoping that if he could not communicate with his other ships he could at least get some news about them from General Banks.

When signal flags hoisted to the top of the *Hartford*'s mast failed to elicit any response Farragut's secretary, Edward C. Gabaudan, volunteered to try to make his way past Port Hudson with orders for the vessels below there and dispatches for the Navy Department.

Soon after dark April 8 Gabaudan and an escaped slave embarked in a skiff covered with branches to make it look as much as possible like a floating tree, a fairly common sight on the Mississippi. Once this craft drifted close enough to the shore for its occupants to hear some Confederate sentries commenting on the hugeness of the log they thought they saw. On another occasion some men rowed out to take a look at the supposed log. Fortunately for Gabaudan, more so for his Negro companion, they made only a cursory examination of it. A two hour voyage, which must have seemed many times longer to the pair, carried them safely past Port Hudson.

Eight days later Farragut returned to the vicinity of Port Hudson and communicated with the ships below there by signalling across Thomas Point from the *Hartford*'s masthead.

Although Porter had been quick to describe the Yazoo Valley expedition, in which he did not participate, as a failure, he did not admit that the Steele's Bayou expedition which he led was also a failure. In his view a mission he could not accomplish was per se an impossible one. Therefore, he concluded that the Navy could not secure a beachhead for the Army anywhere near Vicksburg. He felt sure, as he reported to the Navy Department, that the Confederate batteries in that vicinity could easily destroy four times as many gunboats as there were in the Mississippi Squadron. This being so, the only thing he could see left to do was for an army of 150,000 men to march from Memphis by way of Grenada, Mississippi, prepared to stay in the field until Vicksburg was taken.[11]

Grant, who was as strongly averse now as he had been at the beginning of the year to going back to Memphis, suggested a different plan. He proposed to march most of his troops down the undefended westerly side of the Mississippi to a point well below Vicksburg's southernmost defenses, cross the river with the help and protection of some gunboats, if Porter would send them downstream, and come upon the city from behind. Basically this plan resembled the one successfully used by Pope and Foote at Island No. 10. Grant's plan dif-

11. Ibid., Ser. 1, 24: 479.

fered from Pope's in the much greater distance Grant would have to cover and, most importantly, in the fact that Grant intended deliberately to abandon his base of supplies and live off the country as he had been forced to do after the debacle at Holly Springs.

Porter agreed to send the gunboats *Benton, Lafayette, Louisville, Mound City, Pittsburg, Carondolet, Tuscumbia,* and *General Sterling Price* below Vicksburg to help Grant.

The boats were prepared for the forthcoming ordeal by having the after ends of their casemates, their weakest parts, covered with bales of hay and logs. To have a supply of fuel available each boat except the *Benton* had a barge loaded with 10,000 bushels of coal lashed to her starboard side, leaving their port sides clear. The slow, clumsy *Benton* was accompanied and assisted by a towboat.

The gunboats and three empty transports left their moorings near the mouth of the Yazoo River soon after 9 P.M., April 16, under orders to steam slowly at 50-yard intervals with each boat a little to one side of the one next ahead so that if any of them were disabled those following could pass easily.

Although no lights were shown and precautions were taken against making any avoidable noise or smoke, the boats were discovered soon after 11 P.M. In the ensuing bombardment the *Benton* was hit four times and had four men wounded, one of them seriously. The *Lafayette* sustained nine hits and had the coal barge she was towing sunk, but had no casualties or serious damage. The *General Sterling Price* collided with the *Louisville* and had her superstructure considerably cut up by 13 hits, but suffered no casualties. The *Louisville* lost the barge she was towing, but picked it up again. The *Mound City* was hit six times and had as many men wounded. The seven shots that hit the *Pittsburg* did little damage and caused no casualties. The *Carondolet* had to round to once to avoid a collision and her captain attributed her escape with only two hits and four men slightly wounded to the fact that this "pirouette" disconcerted the Confederate gunners. The transport *Henry Clay,* sunk by a heavy shot, was the only boat lost and her crew was saved. The *Tuscumbia,* last in line, ran aground on the Louisiana side of the river, but freed herself and got through with no casualties.

Despite the severe fire to which they had been subjected all of the boats were fit for further service an hour and a half after they anchored below Vicksburg.

Greatly encouraged by the relatively successful passage of the transports, Grant sent six more of them downstream during the night of April 22-23. One of them was sunk, the others got through safely.

On April 28 Grant and Porter met at Hard Times, Louisiana, about 70 miles below Vicksburg, and made plans for the gunboats to reduce the Confederate works at Grand Gulf to permit the Union troops to cross the river there.

Early the following morning the gunboats closed Grand Gulf, which was now defended by two 8-inch rifles, two 7-inch rifles, a 100-pound rifle, a 30-pound rifle, two 32-pound smoothbores, and five light pieces. Some of the boats ran past the Confederate batteries and rounded to so they could fight with their bow guns; those that stayed upstream used their stern guns. After nearly six hours of hot work the gunboats broke off the engagement with seven men killed, 19 wounded in the *Benton;* five dead, 24 wounded in the *Tuscumbia;* six dead, 13 wounded in the *Pittsburg;* and one wounded in the *Lafayette.*

Unable to waste any time anywhere if his plan was to succeed, Grant arranged to meet Porter again nine miles below Grand Gulf at a point opposite Bruinsburg, Mississippi. The troops marched immediately after Grant and Porter finished their brief council of war. Some hours later (at 8 P.M.) the gunboats bombarded the Confederate works to furnish a cover for the passage of the transports, then followed them past Grand Gulf. At daybreak, April 30, the soldiers began crossing the river with the gunboats as well as the transports serving as ferryboats.

On this same day the *Tyler, Choctaw, Baron de Kalb, Signal, Romeo, Louisville, Petrel, Black Hawk,* three mortar boats, and 10 transports made a feint against Chickasaw Bluffs to prevent the Confederates from sending troops south to reinforce those who were expected to oppose Grant. Although the Union force did not make an actual assault on the bluffs, the boats were roughly handled; the *Choctaw* in particular was hit more than 50 times, fortunately without suffering any casualties.

When the gunboats that had gone below Vicksburg steamed up to Grand Gulf, prepared for a hot battle, they found the place had been evacuated and Grant immediately shifted his base to there.

After Grant marched inland from Grand Gulf to get behind Vicksburg, Farragut returned to the Gulf of Mexico and the Mississippi Squadron pushed up the Red River and reached Alexandria, Louisiana, on May 6. By this time most Confederate property had been moved to Shreveport, Louisiana, where the Union gunboats could not go because of low water.

On returning from the Red River the Mississippi Squadron carried dispatches, patrolled the Mississippi, etc., but did no particularly vigorous fighting until May 22. On that day the gunboats and mortar boats bombarded Vicksburg heavily at the same time that the Army tried to take the city by assault. The Army's attack failed so Grant settled down to besiege the place with the Navy doing whatever the Army requested.

At the earnest request of General Sherman the *Cincinnati* was sent on May 27 to enfilade some bothersome rifle pits on the bank of the Yazoo River. As she rounded to abreast of the position she was to attack she was hit several times below the water line and a shot wrecked her pilothouse at the same time that another damaged her starboard rudder. With five men dead the sinking gunboat was held against the bank long enough for the 14 wounded men in her to be put ashore. When she went down she carried 15 of her crew with her and they were drowned.

On July 4, 1863, Vicksburg was finally surrendered; within a week Port Hudson was taken, and once again, as President Lincoln said, "the father of the waters ran unvexed to the sea."

9

Charleston—1862-63

Except for Richmond there was no place in the South the Union was more eager to capture or the Confederates were more determined to defend than Charleston. Actually the city had no great strategic value. Unlike Richmond, where the Tredegar Iron Works was located, Charleston was not the seat of a major war industry and its comparatively poor railroad connections with the interior kept it from being a highly important destination or point of departure for blockade runners. But it had an enormous psychological value to both sides —in the eyes of northerners it was the place where the rebellion began; to southerners it was where their war for independence started.

Early in 1862 Fox suggested to Du Pont that if the *Monitor* and *Galena* were added to the South Atlantic Squadron it should be possible for the Navy, acting alone, to capture Charleston.[1] (This suggestion is characteristic of the interservice rivalry that persisted throughout the Civil War.) Because the *Monitor* sank on her way to join the squadron and the *Galena* was found badly wanting in defensive strength the Union

1. Robert Means Thompson and Richard Wainwright (eds.), *Confidential Correspondence of Gustavus Vasa Fox* (New York: De Vinne Press, for the Naval History Society, 1918-19), 1: 114-15, 119, 127.

Navy had to content itself with merely blockading Charleston for nearly a year longer.

During this time the Confederates built two ironclads, the *Chicora* and *Palmetto State,* of the now standardized Brooke-Porter design (except that neither of them nor any other southern vessel fought with her decks submerged as the *Virginia* did).

Funds for the construction of the *Chicora* were provided entirely by the government of the Confederate States, but 10 percent (about $30,000) of the cost of the *Palmetto State* was raised, as the *Richmond Whig* reported, "through the efforts and offerings of the women of South Carolina," who donated plate, jewelry, the widow's mite, and even bridal gifts. At the ship's launching, on October 17, 1862, Miss Sue Gelzer, the first contributor to the fund, broke a bottle of rare old wine over her bow and christened her in the name of the patriotic women of the state.[2]

The *Chicora,* which was about 135 feet long, 35 feet in beam, and drew about 12 feet, had armor 4 inches thick over 22 inches of oak and pine. Her battery comprised two 9-inch Dahlgrens and four banded 32-pound rifles firing 60-pound shot. The *Palmetto State,* 150 feet long, 35 feet in beam, with a draught of about 12 feet, had armor 4 inches thick and mounted an 80-pound rifle forward, a 60-pound rifle aft, and two 8-inch smoothbores in broadside.

At the urging of General P. G. T. Beauregard, commanding officer of Charleston's defenses, the *Chicora* and *Palmetto State* undertook to raise the blockade of that city.

Either by luck or design they sortied at a moment when the *Powhatan* and *Candandaigua,* the most for-

---

2. *Richmond Whig,* October 17, 1862, as quoted in F. Moore, *Rebellion Record* (New York: G. P. Putnam; D. Van Nostrand, 1861–67), vol. 6, docs., p. 15.

midable of the vessels usually stationed off Charleston, were coaling at Port Royal, leaving only three warships, the *Housatonic, Ottawa,* and *Unadilla,* and the converted merchantmen *Mercidita, Keystone State, Quaker City, Memphis, Augusta, Stettin,* and *Flag* there.

The *Chicora* and *Palmetto State* crossed Charleston bar at 4:30 A.M., January 31, 1863 on a tide that gave them a scant two feet of water under their keels. With the weather hazy (visibility 100 yards) they closed the *Mercidita* just as she was picking up her mooring after having chased a supposed blockade runner which had turned out to be a Union transport off its course. When a strange craft suddenly appeared close aboard the *Mercidita* all hands were called to quarters and her guns were run out. On receiving an unintelligible reply to her hail the *Mercidita* opened fire, but none of her guns could be brought to bear upon the other ship because she was too close under the quarter. Less than two minutes after the alarm was raised a large shell hit the *Mercidita* and went through three staterooms and the steam drum of a boiler before it exploded against the ship's side where it tore a hole three or four feet in diameter. Nobody was hurt by the blast, but escaping steam killed two men instantly and mortally injured two others. The loss of pressure, of course, stopped the engines. Because the captain believed the ship was in a "sinking and perfectly defenseless condition"[3] an officer was sent to surrender to her attacker, which proved to be the *Palmetto State*. He gave his parole for the *Mercidita*'s officers and crew, but by oversight the Confederate captain did not mention the ship. The *Mercidita*'s captain soon found she was not really in danger of sinking and when no one came on board to take possession

3. *Civil War Naval Chronology* (Washington, D. C.: Naval History Division, Office of the Chief of Naval Operations, Navy Department, n. d.), pt. 3, p. 18. (Hereafter cited as *NC*.)

of her he had her towed away; an act which led the Confederates to make some harsh remarks about Yankee trickery.

After the *Palmetto State*'s victory over the *Mercidita* the Confederate vessels turned their attention to the *Keystone State*. Before the *Mercidita* was surrendered she had burned coston lights to alarm the *Keystone State* and that ship was getting under way when the *Chicora* and *Palmetto State* closed her. Firing ineffectively at her armored attackers at a range of 50 yards, the *Keystone State* evaded an attempt to ram her, but she was set on fire by a bursting shell. Ten minutes later, with the fire extinguished, she headed for one of the Confederate ships at full speed, determined to run her enemy down. Making 12 miles an hour, with all of her guns depressed for a plunging fire at the moment of collision, the *Keystone State* was stopped by a shell that passed through the steam drums of both of her boilers, making a hole four feet long in one of them and a slightly smaller hole in the other. Two more shells burst on the quarter-deck and seven struck the hull at or near the water line. These 10 hits killed 20 of her complement of 196 men, mortally wounded four more, and less severely hurt another 16; most of the injured suffered steam burns. The captain, believing the ship to be helpless, had her colors hauled down, but when the chief engineer reported that she could keep going for a while longer she renewed the fight until she could be taken in tow by the *Memphis*. While the *Keystone State* was being towed to Port Royal she had to run her starboard guns way out and haul her port ones far in to keep heeled far enough over to prevent the holes along her side from causing her to sink.

The *Quaker City* was hit by a shell from one of the Confederate vessels amidships about seven feet above the water line. Part of a guard rail and a guard brace were cut away, considerable damage was done in the engine room, and "sad havoc [was made] with the bulkheads."[4]

The *Augusta* was also slightly damaged by a shell before the engagement ended at about 8 A.M. when the *Chicora* and *Palmetto State* returned to Charleston fully satisfied with the work they had done.

The Confederate government promptly proclaimed to the foreign consuls that the blockade had been raised and would have to be reestablished after 60 days notice at "the now open port of Charleston," the *te deum* was sung at a well attended public celebration held in St. Philip's Church, and, as a local newspaper reported, the countenances of the city's residents beamed more brightly than at any time "since the joyous news was passed . . . that Fort Sumter had yielded to General Beauregard" on April 13, 1861.[5]

The Confederates had reason to rejoice. Some of their ironclads had at last left a harbor to fight at sea and the claim that they had raised the blockade was denied so stridently as to lead one to suspect that the Union ships did retire, at least briefly, to a discreet distance after the encounter with the *Chicora* and *Palmetto State*. However, the *New Ironsides* was sent to Charleston soon afterward and the blockade became stricter than ever.

Early in May, 1863 the "improved monitors" *Weehawken, Passaic, Montauk, Patapsco, Catskill, Nantucket,* and *Nahant,* and the *Keokuk,* an experimental ironclad, joined Du Pont's South Atlantic Squadron.

The improved monitors (of which more than 50 were built during the Civil War) had pilothouses eight inches thick located on top of the turrets where they did not limit the line of fire, their turrets were 11 inches thick,

4. Ibid., pt. 3, p. 19.
5. *Charleston Courier,* as quoted in Moore, *Rebellion Record,* vol. 5, p. 413.

and a few of them had two turrets, enabling them to mount four guns. But they cannot possibly be described as efficient warships.

Their worst fault was their lack of firepower. To improve them in this regard they were supposed to have been armed with 15-inch Dahlgrens. Because it was impossible to obtain enough of such guns in time most of the monitors assigned to Du Pont's squadron carried one 15-inch piece and one of 11 inches. The *Patapsco* had a 15-inch smoothbore and a 150-pound Parrott rifle. (The latter was a dangerously heavy gun because Parrotts firing more than 30-pound projectiles often burst.)

The monitors' pilothouses, eight feet in inside diameter, afforded little space for the three men—captain, pilot, and helmsman—who occupied them and it was impossible to see much through the small circular or oval holes which served in lieu of windows.

The plates of the monitors' pilothouses and turrets were held together by bolts with the heads outside, the nuts inside. Hits by heavy projectiles often sent the nuts flying around in the pilothouses and turrets, gravely endangering their occupants, and the rebound of the plates sometimes withdrew the bolts entirely. Although no pilothouse or turret ever collapsed on this account some of them almost did so.

To supply ammunition to a monitor's guns it was necessary to align a hole in the turret floor with an opening in the deck below the floor. With the hatches battened down, as they had to be in action or in stormy weather, the same holes afforded the only means of egress from below deck.

Although the monitors had an easy motion in heavy seas, they strained the fastenings of their bow, stern, and side overhangs badly, causing them to leak more or less dangerously.

Newly launched monitors had a top speed of a little

more than seven knots; with their bottoms foul, as they soon became in southern waters, they did well to make four knots. Their hull form made them difficult to steer at slow speeds, particularly in shallow water, and at anchor they yawed wildly.

Besides being slow and unwieldy monitors were uncomfortable to live in. It was usually hot and damp below decks. Temperatures of 100 degrees Fahrenheit in the living quarters and 150 degrees in the engine and boiler rooms were not uncommon. At the latter temperature it was impossible to touch metal with bare hands and the men had to be furnished with pieces of canvas similar to the pads housewives used with old-fashioned flatirons.

Anyone who ventured onto a monitor's deck except in the calmest sort of weather risked being washed overboard. Even in a flat calm the deck was a dangerous place to be unless the ship was far enough from shore to be out of range of sharpshooters.

An officer who served in one of these craft wrote: "I will never again go to sea in a monitor. . . . No glory, no promotion can ever pay for it."[6]

Captain Du Pont thought the monitors would have been much better vessels if Ericsson had had a couple of sailors at his elbow when he designed them.[7] However, they were well enough liked by Welles's successors, most of whom were also landsmen, for the last of them to have been kept in active service until 1926 and on the *Navy List* for another 11 years.

The *Keokuk,* built at the same time the improved monitors were, looked like a better fighting craft than they did and was certainly more livable. She was 159 feet long, 39 feet in beam, drew eight feet, and had a freeboard of about five feet—enough to make it reasonably safe on deck. Armored with four inches of iron over

6. *NC,* pt. 5, p. 32.
7. Ibid., pt. 3, p. 44.

four inches of wood, she had two almost circular fixed casemates, each containing a pivot-mounted 9-inch Dahlgren which could be fired through any of three gunports spaced 90 degrees apart. Thus she had almost as much firepower as the *Monitor* and could train her guns through an even bigger arc.

Many northerners confidently believed these supposedly powerful vessels could force their way into Charleston harbor whenever Du Pont chose to have them do so. Welles and Fox were of this opinion and Welles wrote to Du Pont on January 6, 1863: "The ironclads have been ordered to and are now on their way to join your command to enable you to enter the harbor of Charleston and demand the surrender of all its defenses or suffer the consequences of a refusal."[8]

Welles and Fox would have expected less of the monitors if the former, a landsman, had appreciated the tactical situation at Charleston and Fox, a sailor, had not disregarded it. A vessel of a monitor's draught bound for that city had to approach the coast near the southern end of Morris Island, sail close (within a mile) along the island's eastern shore, and make its way through a narrow, winding channel with unbuoyed shoals on either side, past a total of 130 heavy cannon mounted in Battery Wagner (called Fort Wagner in Union reports) and Battery Gregg on Morris Island. Next it would have to pass Fort Sumter on a man-made islet in the middle of the harbor, and Battery Beauregard, Fort Moultrie, and Battery Bee on Sullivan's Island; then it would find itself covered by 17 more big guns in what Du Pont called a *"cul de sac"* comparable to "a porcupine's hide turned outside in . . . with no outlet."[9]

---

8. *Official Records of the Union and Confederate Navies in the War of the Rebellion* (Washington, D. C.: Government Printing Office, 1894–1922), Ser. 1, 13: 503. (Hereafter cited as *ORN.*)

9. *NC,* pt. 2, p. 98.

Because Du Pont wanted to test the monitors before taking a fleet of them into a foreseeably hot battle he ordered the *Montauk,* the first one to reach him, to attack Fort McAllister at Genesis Point on Ossabaw Sound on the Georgia coast. This work, at the mouth of the Great Ogeechee River, was armed with a 10-inch columbiad, an 8-inch columbiad, one 42-pound, and five 32-pound smoothbores.

The *Montauk* arrived at Ossabaw Sound in tow of the *James Adger,* Saturday, January 24, 1863. She crossed the bar at half past five that afternoon, but had to anchor immediately because of a fog which lasted through the following day.

After the fog lifted the *Montauk* opened fire on the fort from a position about 150 yards below a line of piles extending across the Great Ogeechee River. The wooden gunboats *Seneca, Wissahickon, Dawn,* and *C. P. Williams,* which had been blockading the river, lay behind the *Montauk* and fired over her. By noon the *Montauk* had exhausted her ammunition so she and the gunboats withdrew. The monitor had suffered no casualties and no serious damage from the 13 hits she had sustained, but she had not done any harm to the fort.

In the belief that more could have been accomplished at a shorter range the *Montauk*'s captain stood within 600 yards of the fort the following day. During the four hours the ship bombarded the fort she took 48 hits without suffering any casualties or being severely damaged, but again all she did to the Confederate works was to scatter some dust and sand.

As a result of this trial of strength Du Pont asked himself: "If one ironclad [mounting two guns] cannot take eight guns—how are five [monitors with 10 guns] to take 147 guns in Charleston harbor?"[10]

Conversely, the *Montauk*'s inability to damage Fort

10. Ibid., pt. 3, p. 17.

McAllister encouraged the Confederates by showing them, as General Beauregard put it, "that iron-clads were not as formidable as they were supposed to be against land-batteries."[11]

However, the *Montauk* did not entirely waste her time at Fort McAllister. On February 27 she discovered the side-wheeler *Nashville* (she had been originally a blockade runner, later a Confederate privateer named *Rattlesnake*—but had taken no prizes—and was now back in her old trade under her original name) aground in the Great Ogeechee River. The following morning the *Montauk* set the *Nashville* on fire after shelling her for 20 minutes. Then, on her way downstream, the *Montauk* struck a torpedo. She was kept from sinking by being quickly beached and she was repaired without too much trouble at low tide. However, this event led her people and those of the other monitors to do some serious thinking about the almost escape-proof qualities of such vessels.

To test the monitors further and to give their crews practice in handling them Du Pont sent the *Passaic, Patapsco,* and *Nahant* to Fort McAllister a few days later.

Like the *Montauk* they found themselves unable to do any harm to the fort and one of them suffered some fairly severe damage during a six-hour engagement. A 10-inch shell filled with sand hit the *Passaic*'s deck and if it had not landed on a beam it would have gone through the deck, and perhaps the vessel's bottom too. As it was it crushed the planking beside the beam and opened a large hole in the deck. This would have been a serious thing if the weather had been at all rough.

His experiences with the monitors indicated to Du Pont and to his immediate subordinates that because of

11. Robert U. Johnson and Clarence Buel (eds.), *Battles and Leaders of the Civil War* (New York: The Century Co., 1884–87), 4: 8.

their lack of firepower these vessels would not be of much use against fortifications. However, Welles and Fox remained convinced that they were perfectly suited for such work; after all, they had been specifically designed for just that purpose.

Wanting to prove themselves right about the monitors' capabilities and eager for the Navy to outshine the Army, Welles and Fox strongly urged Du Pont to attack Charleston, though they did not flatly order him to do so. In these circumstances Du Pont reluctantly decided to do the best he could with the monitors in order to meet the wishes, virtually amounting to directives, of his superiors, and to fulfill, if possible, the heady expectations of the public. On Saturday, April 4, 1863 he issued the following order to the commanding officers of the *New Ironsides,* the seven monitors, and the *Keokuk:*

> The Squadron will pass up the main ship channel [at Charleston] without returning the fire of the batteries on Morris Island, unless signal should be made to commence action. The ships will open fire on Fort Sumter when within easy range, and will take up a position to the northward and westward of that fortification, engaging its left or northeast face at a distance of from 600 to 800 yards, firing low and aiming at the center embrasure. The commanding officers will instruct their officers and men to carefully avoid wasting shot and will enjoin upon them the necessity of precision rather than rapidity of fire. Each ship will be prepared to render every assistance possible to vessels that may require it. . . . After the reduction of Fort Sumter it is probable that the next point of attack will be the batteries on Morris Island. The order of battle will be line ahead. . . . A squadron of reserve . . . will be formed outside [of] the bar and near the entrance buoy, consisting of the . . . *Canandaigua, Housatonic, Huron, Unadilla,* [and] *Wissahickton,* [which] will be held in readiness to support the ironclads when they attack the batteries on Morris Island.[12]

At daybreak Sunday, the *New Ironsides, Keokuk,* the

12. *NC,* pt. 3, pp. 57–58.

monitors, and the wooden gunboats mentioned in Du Pont's order left North Edisto Island, near Port Royal. Five hours later they reached Charleston bar, about six miles from Fort Sumter. Most of the vessels anchored there while the *Keokuk,* accompanied by the *Patapsco* and *Catskill,* buoyed the bar. When this task had been accomplished the two monitors anchored inside of the bar to guard against removal of the buoys while the others of the force lay outside for the night.

Early Monday morning the fleet crossed the bar prepared to attack Fort Sumter immediately. However, this plan had to be abandoned when the pilots refused to attempt the channel because it had become so hazy that no landmarks could be seen.

The following morning a breeze too light to roughen the sea blew the haze away and Du Pont ordered the ships to get under way at 11 A.M., when they could take advantage of slack water.

As the *Weehawken*'s anchor was being weighed it fouled some grapnels hanging underneath a heavy raft (designed by Ericsson), 50 feet long and shaped like a bootjack, that had been fastened to her bow for the purpose of clearing torpedoes from the channel. This mishap delayed the movement of the fleet for more than two hours and gave Fort Sumter's garrison time to eat a leisurely dinner. Finally the ships got under way in line ahead with the *Weehawken* followed by the *Passaic, Montauk, Patapsco, New Ironsides* (flagship), *Catskill, Nantucket, Nahant* and *Keokuk*. Fort Sumter's drummer beat the long roll at 2:30 P.M. Half an hour later Fort Moultrie opened fire on the *Weehawken.* With her reply the battle began.

From the first things went badly for the ships. The raft the *Weehawken* was pushing made her extremely difficult to steer. Her erratic behavior threw the rest of the column into disorder and the other vessels dodging her and each other could not maintain their prescribed intervals and had a great deal of trouble in holding their positions in line. As they steamed up the channel they came upon buoys scattered in all directions. The suspicion that they might be equipped with torpedoes was quickly confirmed by an explosion near the *Weehawken*. Although no harm was done to the vessel, this event had a sobering effect on the monitors' captains. A little later when the *Weehawken* came upon several rows of anchored casks extending from Fort Sumter to Fort Moultrie she promptly veered to port (toward the south) because her captain "thought it not right to entangle the vessel in obstructions which [he] did not think [she] could have passed through, and in which [she would] have been caught."[13] This unexpected movement confused the other ships' captains and the fleet went into a huddle under a crossfire from all of the guns in Charleston's outer defenses.

The *Weehawken* was hit more than 50 times in less than the same number of minutes. Her armor was badly broken at one place. A shot penetrated her deck and caused a dangerous leak. A shell fragment wedged between the top of the turret and the bottom of the pilothouse temporarily prevented the turret from turning. Many bolts in the turret and pilothouse were broken, gravely weakening them.

The *Passaic,* hit 35 times, had her turret put out of order twice, once after she had fired her 11-inch gun only four times, again after nine shots from her 15-inch piece. All of the plates at the upper edge of her turret were broken and the pilothouse was badly dented.

The *Montauk* took 14 hits without suffering any casualties or severe damage. However, her captain reported, as did others commanding ironclads, that he had

---

13. Ibid., pt. 3, p. 59.

> experienced serious embarrassment in maneuvering . . . in the narrow and uncertain channel, with the limited means of observation afforded from the pilothouse under the rapid and concentrated fire from the forts, the vessels of the fleet close around, . . . and neither compass nor buoys to guide [him].[14]

(At this time no method of correcting a compass in an ironclad was known so they did not carry them. On long passages monitors were towed by wooden ships equipped with compasses; near shore, as at Fort McAllister or Charleston, their pilots relied on landmarks.)

When the *Montauk* turned in the wake of the *Passaic* the *Patapsco* was forced to swing so quickly that she lost headway and steerageway. While she lay, as if hove to, about 600 yards from Fort Moultrie and 1200 yards from Fort Sumter, she was hit 47 times. By luck she suffered no worse damage than having her turret jammed, but she would have been completely disabled if she had remained under the same fire much longer.

The *New Ironsides* was almost unmanageable in the swift tidal current of the narrow channel. She had to be anchored once because she had collided with a couple of monitors and twice more to avoid being swept ashore. Finally Du Pont signalled for the fleet to disregard the flagship's motions and she took practically no part in the battle. At 1000 yards, the closest she came to Fort Sumter, she was hit 50 times, but suffered no casualties or serious damage. It was afterward learned that she had spent much of the afternoon anchored directly over a mine containing a ton of gunpowder which failed to explode, despite repeated attempts to set it off, because a wire leading to it had been broken when a wagon was driven across it.

The *Catskill,* hit 20 times, was undamaged except for a hole through her deck—a thing of no great matter

14. *ORN,* Ser. 1, 14: 14.

because the sea was calm, but enough to have sunk her if it had been rough.

The *Nantucket,* which spent almost an hour within 750 yards of Fort Moultrie, was considerably damaged by 51 hits. After her 15-inch gun had been fired three times the gunport jammed. Later a broken bolt prevented the turret from turning until repairs were made under fire.

The *Nahant* suffered the worst casualties and damage of any of the monitors. An 80-pound piece of armor broken from the pilothouse mortally hurt the quartermaster, slightly injured the captain and the pilot, and temporarily disabled the steering gear. So many bolts were broken in the pilothouse and turret as to lead the captain to believe they both would have collapsed if the ship had stayed under fire for half an hour longer than she did. The man killed and six others hurt in the *Nahant* by flying nuts were the monitors' only casualties.

The *Keokuk* took by far the harshest punishment of any of the fleet. While maneuvering to avoid a collision with a monitor she was forced to spend 30 minutes under a crossfire from Forts Sumter and Moultrie at a range of less than 600 yards from Fort Moultrie. In that time she sustained 90 hits, 19 of them between wind and water. When she drew out of range her forward casemate had been completely wrecked and so many men had been wounded in the after one that its gun could not be served. Calm weather made it possible to keep her afloat overnight, but when a breeze roughened the sea slightly in the morning she sank. By good luck she went down slowly enough to permit all of her people, including the captain and 15 other severely wounded men to be saved.

(Because they were under orders not to act unless the Union ships passed Fort Sumter the *Chicora* and *Palmetto State* took no part in the battle.)

The results of this engagement did not even begin to compensate for the damage the ships suffered. They did little damage to Forts Sumter and Moultrie, none at all to the other works, and the Confederate casualties were only three killed, three severely wounded, and three slightly wounded by gunfire—another man was mortally injured by a falling flagpole.

Du Pont recalled the ships at 5 P.M., fully intending to make another attempt to enter harbor the following day. He changed his mind when his subordinates told him that five of the monitors were practically unable to fire their guns and that if the battle had lasted half an hour longer the other two would have been in equally bad shape. These reports convinced him "that a renewal of the attack could not result in the capture of Charleston, but would, in all probability, end in the destruction of a portion of the ironclad fleet and might leave several of them sunk within reach of [the Confederates]."[15] To carry on would, he thought, result only in converting a failure into a disaster. Therefore, as he reported to the Secretary of the Navy, he "determined not to renew the attack."[16]

He made this decision without consulting any of his subordinates because no one else's opinion would have changed his. However, he learned later that if he had talked to the monitors' captains they would all have concurred with him.

Although most reports sent to the War and Navy Departments were released to the newspapers almost as soon as they were received, Du Pont's reports about his attack on Charleston were kept secret by Welles and Fox, who were unable to believe—or unwilling to admit—that they had overrated the monitors. Thus the public was not allowed to learn that the attack had failed as

15. *NC,* pt. 3, p. 61.
16. Ibid.

a result of weakness of the monitors' cast iron and wrought iron parts, not from lack of Du Pont's courage or that of any of his subordinates. Worse yet, Welles and Fox, in desperate need of a scapegoat, let it become widely known that they thought Du Pont had been much too ready to give up the fight. Their attitude led Du Pont to suggest that he be relieved if the Navy Department supposed that someone else could do more than he had. Taking him at his word the Department named Captain Foote to succeed him.

News that Foote, who was known to have great faith in ironclads, had been ordered to command the South Atlantic Squadron led the public to expect Charleston to be taken quickly despite the setback suffered by Du Pont.

There is absolutely no reason to imagine that the Navy could have done any more at Charleston under Foote, without substantial help from the Army, than it had done under Du Pont. However, what Foote might have accomplished will always remain a matter of speculation. While he was on his way to Charleston he died from the effects of the wound he had suffered at Fort Donelson.

Admiral Dahlgren, who was chosen to succeed Du Pont after Foote died, had spent nearly 20 years on shore as chief of the Bureau of Ordnance. He had gained well deserved fame in that position, but, eager to make a name for himself as a seagoing officer, he began angling for the command of the South Atlantic Squadron the moment he had any inkling it might be attainable. He finally got the place through President Lincoln's influence. Dahlgren's presence in Washington had enabled him to keep in close touch with the President, who was keenly interested in the "tools of war"; his loyalty during the critical early months of the war when many other naval officers "went south" won Lincoln's confidence and the President practically, if not explicitly, told Welles to put Dahlgren in command of the squadron.[17]

---

17. Gideon Welles, *Diary of Gideon Welles* (Boston: Houghton, Mifflin and Company, 1911), 1: 158, 163–65, 239, 312, 341.

10

Charleston—1863-65

Before Dahlgren reached Charleston, Du Pont had the satisfaction of reporting the capture by two of his ships of the Confederate ironclad *Atlanta,* formerly the *Fingal,* a blockade runner built in England.

The *Fingal,* the first vessel to run the blockade for the account of the Confederate government, had arrived at Savannah on a foggy night in mid-November, 1861, carrying the largest military cargo ever brought into the Confederate States—10,000 British Enfield rifles, 1 million ball cartridges, 2 million percussion caps, 3000 cavalry sabers, 1000 carbines, 1 million rounds of carbine ammunition, 1000 bayonets, 500 revolvers, two large rifled cannon, two smaller rifles, 400 barrels of gunpowder, medical supplies, and cloth for making uniforms. (This suggests how much might have been accomplished if the Confederate government had engaged more extensively in blockade running on its own account instead of leaving it almost entirely to be conducted by "free enterprise," or for the sake of profit without regard to the needs of the country.)

Before the *Fingal* was ready to depart from Savannah the port was too strictly blockaded for her to get out so the Confederate government bought her for conversion into a warship.

The *Atlanta,* 204 feet long, 41 feet in beam, with a draught of 15 feet, 5 inches, had a casemate, similar to the *Virginia*'s, made of 4 inches of iron backed by 3 inches of oak and 15 inches of pine. Her armament included two 6.4-inch, 32-pound rifles mounted in broadside and two 7-inch rifles, pivoted to permit them to be used either in broadside or as bow and stern chasers. She also had a percussion torpedo attached to a spar extending from her bow, which, her captain said, he knew would work to his entire satisfaction if he could touch a Union vessel with it. The *Fingal*'s top speed had been 12 or 13 knots; the *Atlanta,* carrying a heavy weight of armor and armament, could not make more than eight knots, but that was fast enough to overtake any monitor.

The Confederates, as unduly optimistic about the *Atlanta* as many northerners were about the monitors, expected their new ironclad to disperse the blockaders at Savannah, Charleston, and Wilmington, then to carry the war up the Atlantic coast from Delaware to Maine.

Information furnished by Union spies and Confederate deserters kept Du Pont fully aware of the progress being made in converting the *Fingal* into the *Atlanta.* When the work was nearly done he sent the monitors *Nahant* and *Weehawken* to Wassaw Sound to watch for the Confederate vessel to try to make her way to sea.

Leaving Savannah late in the afternoon of June 15, 1863, the *Atlanta* steamed down the Wilmington River, anchored at 8 P.M., and spent some time coaling. This task finished, she dropped downstream to within five or six miles of the monitors, but was concealed from their view, ready to move on them at early dawn.

Soon after 4 A.M. June 17, a lookout in one of the monitors sighted the *Atlanta* coming down the Wilmington River accompanied by two steamers filled with sightseers eager to witness her expected triumph.

At first the Confederates had reason to feel delighted

with events. The monitors no sooner saw the *Atlanta* than they headed toward the southeastern or ocean end of Wassaw Sound. Their captains had learned enough at Charleston about the difficulty of handling them in narrow waters to cause them to seek sea room if it was available, but the sightseers and the *Atlanta*'s captain thought the monitors were trying to get away. The *Atlanta* went after them at full speed, hoping to torpedo one of them and confident of defeating both of them. When they turned around to join battle at 5 A.M., the *Atlanta* opened fire on the *Weehawken* at a range of a mile and a half and scored a complete miss. Soon afterward the *Atlanta* had the misfortune to run aground. With the tide rising she freed herself in 15 minutes, but the current made her hard to steer and she soon grounded again, this time hard and fast.

Because the *Nahant* had no pilot she followed in the wake of the *Weehawken* which closed the range to 300 yards before she opened fire at 5:15 A.M. At that distance a 15-inch cored shot broke the armor and wooden backing of the *Atlanta*'s casemate, but did not penetrate it. However, the concussion knocked down 40 men, and 16 more were hurt by iron and wooden splinters. An 11-inch shot hit the ship near the water line without doing her any serious damage. Another 15-inch shot took off the top of the pilothouse and wounded two of its three occupants. A third 15-inch shot broke a gunport. The last shot fired by the *Weehawken* missed the *Atlanta* completely.

The *Atlanta,* unable to move, could hardly bring a gun to bear upon her attacker. She fired only six times before her colors were struck and her officers and men surrendered and the disappointed sightseers went back to Savannah.

This affair lasted so short a time—scarcely half an hour—that the *Nahant* did not get into action at all.

However, she collided with the *Weehawken* and started a few of the latter's plates after the fight ended.

Just before Dahlgren relieved Du Pont, General Quincy A. Gillmore, who had recently been put in command of the Union troops deployed against Charleston, asked for naval cooperation in a movement against Battery Wagner. In view of Dahlgren's imminent arrival Du Pont left it for Dahlgren and Gillmore to make their own plans.

Gillmore arranged with Dahlgren to have the monitors *Catskill, Montauk, Nahant,* and *Weehawken* cover the troops as they crossed Lighthouse Inlet, between the northern end of Folly Island and the southern end of Morris Island, and to have the ships bombard Battery Wagner at the same time that the soldiers attacked it. The monitors were ready to play their assigned part in this operation on July 8, 1863, the time set by Gillmore for its occurrence; but he postponed it from day to day until July 10. At 4 A.M. that day his artillery opened fire and the monitors crossed the bar at the same time. By 9 A.M. the troops, using small boats mostly furnished by the Navy and armed with Navy howitzers, had landed on Morris Island. With the monitors firing a rolling barrage ahead of them the soldiers made their way across a range of sand dunes and advanced along the low, flat ground toward the battery nearly two miles away. About 10 A.M. the monitors anchored abreast of the battery at a distance of 1200 yards—the closest the tide would permit them to get—and briskly exchanged shots with the Confederates until noon. After steaming out of range long enough for their crews to eat lunch the monitors bombarded the battery again until 6 P.M.

In this engagement the *Weehawken* was entirely untouched, the *Montauk* was hit twice, and the *Nahant* three times. But, as Dahlgren remarked, the Confederates "seemed to have made a mark of the *Catskill,*"[1] perhaps because she was the flagship. She sustained 60 hits, including 17 on the turret and three on the pilot-house. Dahlgren narrowly escaped being hurt, if not killed, by a flying nut and the casualties included a case of concussion of the brain suffered by a man who was leaning against the turret wall when a shot struck it. There were also five cases of heat prostration among the engine room crew. And, as things turned out, the whole affair was of no avail. When the troops stormed Battery Wagner at daybreak, July 12, they were thrown back almost to their beachhead.

Gillmore decided to cover his next assault with some light artillery and Dahlgren agreed to aid the troops with all of the ships he had available. This movement, scheduled for July 16, was postponed to July 18 because of a storm and it took most of that morning for the Army artillerists to dry their gunpowder.

As soon as the soldiers were ready to move the *New Ironsides, Montauk, Catskill, Nantucket, Weehawken, Patapsco* and the gunboats *Paul Jones, Ottawa, Seneca, Chippewa,* and *Wissahickon* anchored abreast of Battery Wagner. The wooden gunboats lay behind the ironclads and used their rifled pivot guns to fire over them. With the tide ebbing when the bombardment began at noon the pilots did not deem it prudent to take the ironclads nearer shore than the inner edge of the Main Ship Channel, about 1200 yards from their target. As the tide rose the range was gradually closed to 300 yards and the battery fell silent under the ships' heavy fire. However, the Confederate casualties were light—eight men killed, 20 wounded—and the battery sustained no damage at all.

When the ships ceased fire because it had grown too

1. *Civil War Naval Chronology* (Washington, D. C.: Naval History Division, Office of the Chief of Naval Operations, Navy Department, n. d.), pt. 3, p. 113. (Hereafter referred to as *NC.*)

dark to distinguish friend from foe the troops were still 1000 yards from the battery. A few hours later Dahlgren learned that the soldiers had been beaten back with a loss of more than 900 men out of the four regiments most heavily involved.

The Confederates, warned of what was afoot by the ships' bombardment, had amply prepared to meet the assault, even to building rifle pits outside of their works.

In his report about the Navy's part in this affair Dahlgren said, "I have . . . silenced Fort Wagner and driven its garrison to shelter, and can repeat the same, but this is the full extent to which [naval] artillery can go; the rest can only be accomplished by troops."[2] (Du Pont had been openly charged with lethargy and implicitly with cowardice for having hinted at what Dahlgren said openly. For some reason, however, Dahlgren was not even gently chided. Perhaps Welles and Fox did not think it would be wise to treat someone so obviously blessed by the President as Dahlgren had been too harshly; perhaps their need for a scapegoat had been satisfied by the sacrifice of Du Pont.)

Six days later the *New Ironsides, Weehawken, Patapsco, Nahant, Montauk, Paul Jones, Ottawa, Seneca,* and *Dai Ching* again bombarded Battery Wagner and Dahlgren reported another success, saying he had silenced the guns and driven the garrison to shelter.

Despite Dahlgren's optimistic reports Gillmore knew all too well that Battery Wagner could stymie his troops just as long as the Confederates could hold it. Accordingly Gillmore sought Dahlgren's help again. On August 17 the Army's cannon fired over Battery Wagner at Fort Sumter while the *New Ironsides, Weehawken, Catskill, Nahant, Montauk, Patapsco,* and the gunboats *Canandaigua, Mahaska, Cimmaron, Ottawa, Wissa-*

2. Ibid., pt. 3, p. 119.

hickon, Dai Ching, Seneca, and *Lodona* bombarded both the battery and the fort. The Confederates replied with every sort of projectile from minié balls to 10-inch solid shot. At noon, with only Battery Gregg still firing, the ships withdrew because their crews, who had been hard at work since daybreak, needed to rest. The fleet might better have stayed at anchor in the first place for it did no serious damage to any of the Confederate works while Dahlgren's chief of staff, Captain George W. Rodgers (who had been given permission to command his old ship for the day), and another officer were killed in the *Catskill.*

A few days later the Confederates made an attempt to raise the blockade of Charleston and to cause a letup in the constant bombardment of Fort Sumter and Battery Wagner.

About 10 P.M. August 21, 1863, a torpedo boat left Charleston bent on attacking the *New Ironsides*. A couple of hours later the boat closed her chosen prey, the boom was lowered, the engine stopped, and the helm was put over to head directly toward the ship. Nobody in the *New Ironsides* was aware of any danger, but the tide swung her far enough to be missed by the torpedo as the boat drifted past. At this moment a Union lookout discovered the boat, but she was not hit by any of the shots fired at her, although some of them landed at a spot she had just left.

The Confederate captain thought of trying his luck against one of the monitors, but the boat's engine had already given some trouble and his doubts about its reliability led him to return to the city.

During the following night five monitors attacked Fort Sumter at a range of about 1000 yards. The ships got under way before midnight, but hazy weather forced them to steam so slowly that it was 3 A.M. before they were in a position to open fire. An hour later the fog became so thick they could no longer see to shoot. At daybreak they retired without having accomplished anything to justify their having come out.

In his report of this engagement Dahlgren mentioned the fact that only one monitor had gone aground and she had gotten free by her own efforts. He also said he intended as soon as the weather improved to push past Fort Sumter to the inner harbor if the obstructions were not of such a nature as to prevent it.

Three days later Dahlgren ordered the monitors to be made ready to force their way through the obstructions. If he ever meant for his subordinates to take this order seriously he soon changed his mind. After postponing the movement from day to day he tacitly let the order lapse.

On September 1 Dahlgren issued new orders for a night attack on Fort Sumter, but he strictly enjoined the ships to avoid becoming entangled in the obstructions. The monitors fired more than 200 rounds at the fort at a range of 500 yards, then withdrew before daylight, having done nothing to justify their coming out.

When Dahlgren learned that Battery Wagner had been abandoned during the night of September 6-7 he called upon Fort Sumter to surrender. General Beauregard told Dahlgren "to take it if he could."[3]

After receiving this defiant reply Dahlgren sent the *Weehawken* to search for a channel between Cumming's Point and Fort Sumter while the other monitors were ordered to test the obstructions north of the fort. The ships made no serious attempt to pass the obstructions and their gunfire, directed chiefly at Fort Moultrie, did little damage and inflicted only a few casualties.

On the way back to her anchorage the *Weehawken*,

3. Ibid., pt. 3, p. 137.

steering badly because of shoal water, ran aground. Although her coal and ammunition were unloaded into a tugboat, the monitor was not refloated on either of the next two high tides. As luck had it the Confederates did not realize the *Weehawken*'s plight until the morning of September 8—to avoid attracting attention, she had not fired on any of the nearby works. When Fort Moultrie opened fire on her she retaliated by blowing up the fort's magazine, killing 16 men and wounding 12 others.

By now Fort Sumter's walls had been badly damaged and most of its heavy artillery had been removed to other installations. However, Dahlgren thought it was still strong enough to prevent his ships from clearing away the channel obstructions. In this belief he decided to send a landing party against the fort on the night of September 8. On seeking to borrow some rowboats from Gillmore, Dahlgren learned that the General was planning a similar operation for the same night.

After arguing all day and well into the night about whether an Army or a Navy officer should command the movement Dahlgren and Gillmore patched up an agreement calculated simply to prevent a mixup of their separate groups.

Although Dahlgren claimed the naval expedition had been carefully planned and organized, the truth is that it was gotten up on the spur of the moment and prepared with such haste that the man designated to head it, Commander Thomas H. Stevens, first heard of it only after he had eaten his evening meal at the end of a busy day spent getting his ship, the *Weehawken,* afloat again.

Stevens respectfully asked permission to decline the assignment to command the landing party. In his estimation nobody in the fleet had any reliable knowledge of the fort's real condition, not enough time had been taken for the proper organization of a force for such a desperate venture, the gathering of small boats around the flagship during the day would have put the Confederates on their guard, and even if the fort could be taken it would be impossible to hold it.

Without replying categorically to Stevens's objections, Dahlgren said to him, "There is nothing but a corporal's guard in the fort, and all we have to do is to go in and take possession" of it.[4]

Feeling that this answer left him with no choice, Stevens reluctantly undertook the duty. Hoping for the best he hurriedly concocted a plan of action with his two immediate subordinates, but he was not allowed time enough to brief anybody else connected with the expedition. At least one boat set out with no one in it knowing anything about the work ahead of it except that an attack was to be made on something somewhere.

The landing party's boats were loaded and in the water ready to be towed most of the way to Fort Sumter by 10:30 P.M., but the tug did not come for them until after midnight because it took until then for Dahlgren and Gillmore to straighten out their overlapping plans. When the tug finally did arrive its behavior simply added to the existing confusion. As it neared the fort it made three complete circles—a maneuver understood by nobody in the boats—then, without a word of warning, the boats were cast loose while they were heading in a direction which made it extremely difficult, because of the swift flowing tide, to form the three divisions prescribed in Stevens's orders. His plan, which was understood by the division commanders but not by the different boats' captains, was for one division to make a strong feint against the fort's northwest face while the other two divisions made the real attack. As soon as the first division started toward the fort most of the other boats followed it, under the impression that a general movement had been

4. Robert U. Johnson and Clarence Buel (eds.), *Battles and Leaders of the Civil War* (New York: The Century Co., 1884–87), 4: 49.

ordered. Thus, instead of making separate landings all of the boats attempted to land at once. They were met with musketry, hand grenades, grapeshot, and canister while a signal from the fort brought a rain of cannon fire from the batteries on Sullivans and James Islands and the *Chicora* and *Palmetto State*.

As it happened the Confederates had recovered the *Keokuk*'s signal book so they had been able to read the messages exchanged by Dahlgren and Gillmore during the day.[5]

Half an hour after the attack began the boats that had not already landed withdrew, most of them without specific orders to do so, simply because it was the only sensible course to follow in the circumstances. They left four dead men and more than 100 captives behind them.

After this fiasco there was a period of comparative inactivity on the part of the Union forces, but the Confederates kept trying to raise the blockade.

One plan for accomplishing this end, suggested by a high ranking Army officer was to send a ship loaded with gunpowder out against the Union fleet. "Should this explode close to the [*New*] *Ironsides,* or other vessel, the effect must be to destroy her; and if two or three are in juxtaposition, the two or three may be got rid of," he wrote to General Beauregard in mid-July, 1863.[6]

Beauregard rejected this idea on the advice of some naval officers who thought there was little chance that it would do anything but make a lot of noise. However, an attack was made on the *New Ironsides* on October 5, 1863 by the torpedo boat *David,* commanded by Lieutenant W. T. Glassell.

The noise made by the *David*'s engine led to her being discovered while she was about 300 yards from the *New*

5. *NC,* pt. 3, pp. 138–40.
6. Ibid., pt. 3, p. 119.

Ironsides, but Glassell, sitting on top of the low conning tower, steering with his feet, and holding a double barrel shotgun across his lap, disregarded repeated hails and pushed on at full speed toward the ship. At 40 yards range he fired both barrels of the shotgun at the *New Ironsides'* officer of the deck. Almost before the echo of the shots died away the *David*'s torpedo, containing about 60 pounds of gunpowder, was detonated six feet below the ship's water line. The explosion shook the ship heavily and threw up an immense column of water. As the water fell back it nearly swamped the *David* and did put out her fire. Glassell, the fireman, and the engineer, who believed the boat was sinking, all jumped overboard. Glassell was picked up by a Union transport. The fireman clung to the *New Ironsides'* anchor chain until he was taken on board the ship. The engineer started to swim to Fort Sumter, then seeing the *David* still afloat he went back to her and found that the pilot, who could not swim, had never left her. The two men rekindled the fire and made their way back to Charleston in the *David.*

The Confederates were much disappointed when daylight revealed the *New Ironsides* at her usual anchorage. The charge in the torpedo had not been heavy enough to sink her, though she did have to go to Port Royal for extensive repairs.

Mallory said of the attack on the *New Ironsides,* "The annals of naval warfare record few enterprises which exhibit more strikingly than this of Lieutenant Glassell the highest qualities of a sea officer."[7]

The routine of the blockade was interrupted again by the sinking of the *Weehawken* at her mooring inside of Charleston bar on December 6, 1863, a few days after she had taken on as many shells as she could carry. (Because the capacity of the monitors' shell rooms had been

7. Ibid., pt. 3, p. 144.

found insufficient for long continued operations it had become standard practice to store extra shells in their forward holds under decks made with moveable sections. The shells stowed there were easily gotten up when they were needed, but their weight made the vessels lie low in the water and reduced the difference of freeboard between their bows and sterns from its normal 18 inches to six or eight inches. Thus water collected at their bows was slow to run aft to where their pumps were located.) With the morning clear and the sea not particularly rough nobody had bothered to drive the usual rope gasket around the anchor chain to prevent water from leaking through the hawsehole. After a considerable amount of water had slopped into the windlass room the door was dogged shut to keep the cabin dry, but even now no gasket was put around the chain. About 1 P.M. the appearance of water in the cabin caused an officer to realize that the *Weehawken* was in some danger; actually she was doomed by this time. Five minutes after she signalled for help she rolled to starboard, started down by the head, came to an even keel for a moment, then dropped to the bottom like a stone. Twenty-four men were lost with her. Some of them were unable to reach the deck against the inrush of water; others, who had been on deck, could not keep afloat for the few minutes it took for help to reach them.

In August 1863, General Beauregard arranged to have the submarine C.S.S. *H. L. Hunley* shipped on two flatcars from Mobile to Charleston because he regarded her "as the most formidable engine of war for the defense of [the city] at his disposition."[8]

This craft, fashioned from a cylindrical iron boiler with tapered ends attached to it, measured 40 feet long

8. Ibid., pt. 3, p. 128.

and 4 feet in diameter. It had ballast tanks (in the tapered ends) which could be flooded to submerge her by opening valves and emptied to refloat her by hand pumps. Iron weights fastened to the under side of the hull were meant to be dropped by unscrewing bolts extending into the vessel in case extra buoyancy should be needed in an emergency. It was equipped with a mercury depth gauge and a compass for steering when submerged. A candle served to give light and, when its flame began to die, to warn its occupants of the need for fresh air. Windows in the coamings of two hatches, which resembled low conning towers, afforded a view of the outside world when they were above water, but there was no periscope. She was designed to tow a cylindrical torpedo on a line about 200 feet long, dive under her target, surface on the other side of it, and leave the scene after the charge exploded against an enemy ship's hull. The crew consisted of a captain, who was his own helmsman, and eight men who sat facing each other while they turned by hand a crankshaft connected to a propeller which drove the vessel at about four knots.

When the *H. L. Hunley* reached Charleston she was stationed at Beach Inlet on Sullivan's Island to permit her crew of volunteers from the *Chicora* to practice handling her.

On her return to her base on August 29 she was tied up alongside of a steamer. An unexpected move by the latter vessel canted the *H. L. Hunley* far enough to cause her to fill with water and sink with at least five men on board.

After this catastrophe H. L. Hunley, the submarine's designer for whom she was named, asked General Beauregard to have her put under his charge. "I propose," Hunley wrote to Beauregard on September 19, "if you will place the boat in my hands to furnish a crew (in whole or in part) from Mobile who are well acquainted with its management & make the attempt to destroy a vessel of the enemy as early as practicable."[9]

A crew was brought from Mobile and the boat was turned over to Hunley. However, this presumably expert crew proved no more able to handle the *H. L. Hunley* than others had been. On October 15 she failed to surface after a practice dive in nine fathoms of water. It was later learned that she had become stuck in a muddy bottom and the weights attached to her hull could not be released before the air supply was exhausted and her people asphyxiated.

When the *H. L. Hunley* was raised and reconditioned again she was ordered not to do any more diving. She was ballasted to float with her hatch coamings just above water and was equipped with a spar torpedo containing 90 pounds of gunpowder. This device, fitted with a barbed spike, was meant to be fastened to a target ship and exploded by means of a trip line after the boat had backed a safe distance away.

During the next few months the *H. L. Hunley* put to sea on an average of four nights a week, steering on bearings taken from the beach. However, adverse tides or winds repeatedly exhausted the crew before she could reach the nearest of the Union ships, usually anchored six or seven miles from shore.

Finally, on February 17, 1864, the *H. L. Hunley* had the good fortune to come upon the *Housatonic* in the North Channel only about two miles out. Despite the bright moonlight the *H. L. Hunley*'s silent approach enabled her to get within 100 yards of the ship before being discovered. When she was seen the *Housatonic*'s officer of the deck gave orders to slip the anchor chain, back the engine, call the captain, and beat to quarters. Two minutes later, as the ship began to gather sternway, an explosion abaft the mizzenmast blew most of her

9. Ibid., pt. 3, p. 141.

stern away. Settling fast in 28 feet of water she lurched to port and lost the small boats on that side. Some men jumped overboard and were picked up by two boats launched from the starboard side; others saved themselves by climbing into the rigging; because of the call to quarters only five were lost. However, the sinking of the ship was not noticed by the other Union vessels and her survivors were not rescued until after daybreak.

If the *Housatonic* had started ahead instead of backing she probably would have escaped undamaged. As it was the Union and Confederate Navies each lost a vessel for the *H. L. Hunley* never returned to port. She was either destroyed by the explosion of the torpedo or swamped by the wave it caused. But she had written a fateful page in history for her deed foretold the great damage other submarines would do in years to come.

Charleston lost what little strategic importance it had ever possessed when the Confederates abandoned Battery Wagner in September 1863 because control of Morris Island enabled the Union Army to mount enough cannon there to control the Main Ship Channel. However, a number of ironclads, which could advantageously have been used elsewhere, were kept on duty there because Welles and Fox were so obsessed with the place.

With the Confederates trying to drive the Union ships away Charleston remained a fairly busy spot. Early in March 1864 a Confederate torpedo boat attacked the U.S.S. *Memphis*. Although the boat was spotted 50 yards from her target and brought under heavy musketry, she held her course and thrust her spar torpedo, containing 95 pounds of gunpowder, squarely against the side of the *Memphis*. When it failed to explode the Confederate captain coolly tried his luck on the ship's other side. Again the torpedo struck home, but the blow was a glancing one and no attempt was made to detonate the torpedo.

The boat retreated under a hail of gunfire which, to her good luck, missed her entirely.

The same captain and the same boat tried again to sink the *Memphis* about three weeks later (on April 19). This time the boat was sighted at a distance of 150 yards and as the ship got under way she kicked up such a wake that the boat had to turn away to avoid being swamped.

Early in 1865 Dahlgren and Gillmore were ordered, at General William T. Sherman's request, to make a strong demonstration before Charleston to draw the Confederates' attention away from the General's march from Georgia to the north. Before he undertook such a movement Dahlgren wanted the obstructions in the channel located and marked. On January 15, while engaged in this work, the *Patapsco* struck a torpedo. The explosion, under the port bow, caused the vessel to sink in 15 seconds in 50 feet of water. Three men who somehow made their way from below, two who were on top of the turret, and 45 who were on deck survived; 64 others (more than half of the crew) went down with the ship.

Three days later Charleston was evacuated by the Confederates because the presence of Sherman's army in its rear made it untenable, but the city has the distinction of being the only southern seaport to have withstood a determined attack by the Union Army and Navy almost until the end of the war.

11
Albemarle Sound

A little more than a month after the capture of Roanoke Island and the destruction of the Confederate flotilla at Elizabeth City the U.S.S. *Delaware, Stars and Stripes, Louisiana, Hetzel, Commodore Perry, Valley City, Underwriter, Hunchback, Southfield, Morse, Brincker, Commodore Barney,* and *Lockwood,* accompanied by 12,000 troops in transports, set out for New Bern (sometimes spelled Newbern or Newberne), North Carolina, on the left bank of the Neuse River.

Leaving Hatteras Inlet early Wednesday, March 12, 1862, this force anchored shortly before dark the same day at Slocum's Neck, on the right bank of the river, about 15 miles below New Bern, where it was planned to land the troops.

At 6:30 A.M. Thursday, the gunboats shelled the woods near the proposed landing place. Three hours later those soldiers who had disembarked by that time, a number of marines, and three Navy howitzers with their crews started up the river with the gunboats keeping pace with them and firing into the woods ahead of them. The Confederates offered no resistance to this movement until about 4 P.M., when a battery opened fire on the gunboats at long range. As if by tacit consent both sides ceased fire at sunset and the vessels anchored, one behind the other, where they could best protect the men on shore.

Friday morning the soldiers, closely supported by the gunboats, overcame the slight resistance offered by the Confederates at Forts Dixie, Ellis, Thompson, and Lane along the river below New Bern. During the afternoon the troops were ferried across the river and by nightfall they were in full possession of the town.

Leaving a garrison to hold New Bern securely, the combined force turned its attention to Fort Macon, a coast defense installation which guarded the approaches to Morehead City and Beaufort, North Carolina, where the Navy wanted to establish a base. The *Daylight, State of Georgia, Chippewa,* and *Gemsbok* bombarded the fort for a couple of days before bad weather drove them away. However, they had softened up the work enough to enable the troops to take it by assault on April 26.

A short time later Washington on the Pamlico River and Plymouth on the Roanoke River were occupied and garrisoned and half a dozen gunboats were stationed in the North Carolina Sounds as a precautionary measure, but there were not enough troops available for these footholds to be expanded. Because the Confederates were also kept busy elsewhere at this time the area remained pretty quiet until March 14, 1863, the anniversary of the capture of New Bern.

On that date a large Confederate force attacked Fort Anderson, an unfinished work garrisoned by about 300 men across the Neuse River from New Bern. Apparently the Confederates had learned of a request for *all* of the gunboats to cooperate with an expedition to Hyde County and believed that this had been done. Actually the *Hunchback, Ceres,* and *Shawsheen* were nearby. Together with a revenue cutter and an armed schooner, these vessels went quickly to the aid of the fort and drove off the Confederates.

When the Confederates began a siege of Washington about three weeks later (on March 31) two gunboats were left to guard New Bern and one remained at Plymouth while all of the others were sent to help the beleaguered garrison. Their heavy gunfire and the ability of their small boats to carry desperately needed supplies and ammunition to the town finally caused the Confederates to withdraw on April 18.

During the rest of 1863 the area remained so quiet that the officers and men of the Union Army and Navy became somewhat slack. Thus it happened that when they learned there was a Confederate ironclad ram being built at Edwards Ferry on the Roanoke River neither service made any effort to interfere with the work because each thought the other should take care of the matter, even though, as Welles wrote to the Secretary of War, the Union's "possession of the [North Carolina] sounds would be jeoparded" if she got into them.[1]

The *Albermarle,* as the Confederate vessel was named, closely resembled the *Virginia,* but was smaller than her prototype. The *Albermarle* measured 152 feet long, 34 feet in beam, and drew 9 feet of water. Her casemate, about 60 feet long, contained two 100-pound rifles mounted in such a way that they could be used either as bow and stern chasers or in broadside. Her keel was laid in April 1863 and she was launched without her armor, armament, or engines the following November. Once afloat she was towed upstream to be completed at Halifax, North Carolina.

Like all of the Confederate ironclads, the *Albermarle* was built under difficult conditions. With a corn field serving for a shipyard, much of the iron for her construction was obtained by commandeering scrap metal,

1. *Civil War Naval Chronology* (Washington, D. C.: Naval History Division, Office of the Chief of Naval Operations, Navy Department, n. d.), pt. 3, p. 140. (Hereafter cited as *NC.*)

nuts, bolts, plows, tools, flatirons, etc., from every farm for miles around and her builders had no power tools available.

Not surprisingly the *Albermarle* had some weak points. The worst of them were that her draught was too great to permit her to navigate the North Carolina sounds conveniently. She also lacked buoyancy and her decks were so near the water as to make it easy for an enemy vessel to run upon them and any great weight on them threatened to sink her. Nevertheless, she proved to be a formidable opponent of the Union vessels in the sounds.

In mid-April 1864, General Robert F. Hoke of the Confederate Army asked if the *Albermarle* could assist him in an effort he was about to make to retake Plymouth. At this time the *Albermarle* was still having the finishing touches put on her, but the work was quickly completed and she went to Hoke's aid.

On April 18 the *Albermarle* started downstream, going sternforemost and dragging an anchor because the current made her hard to steer. Despite trouble with a propeller shaft and a broken rudder head she anchored 12 miles above Plymouth at 10 P.M. that night.

About 4 A.M. the following day the *Albermarle* headed for Plymouth. She was soon discovered by a Union picket boat which alerted the *Miami,* an eight-gun sidewheeler, and the *Southfield,* a former ferryboat armed with seven guns.

On this occasion the Union Navy was at least not caught utterly by surprise as it had been by the *Virginia* two years earlier. However, experience appears to have taught some high ranking naval officers little about the offensive and defensive qualities of ironclads. At the suggestion of Admiral S. P. Lee, commanding the North Atlantic Squadron, the *Southfield* and *Miami* were connected to each other with a heavy chain to enable them to catch the *Albermarle* between them and hold her there while they pounded her to pieces with their heavy guns.[2]

The merits of this plan were not tested. Instead of coming down the middle of the river and falling into the trap prepared for her the *Albermarle* hugged the shore until she was abreast of the Union vessels, then she cut across the *Miami*'s bow and rammed the *Southfield* hard enough to tear "a hole clear through to her boiler."[3] The *Southfield* sank so fast that the *Albermarle* was unable to back clear quickly enough to avoid being dragged down with the Union vessel until it struck bottom, rolled on its side, and released the ram.

The *Miami* immediately dropped the chain connecting her to the *Southfield* and attacked the *Albermarle*. A shell fired from the *Miami*'s bow gun rebounded from the ironclad's casemate and landed at the feet of Lieutenant C. W. Flusser, the *Miami*'s commanding officer. Its explosion killed him instantly and wounded several other men. (Because the gunboats had been operating against troops they were using shells instead of the solid shot that would have been appropriate for fighting an ironclad.)

A few minutes later the *Southfield*'s captain, who had been rescued by the *Miami* and was now the senior officer present, ordered the *Miami* to withdraw since it would obviously be suicidal for her to oppose the *Albermarle*. This turn of events gave the Confederates the advantage of naval superiority such as Union forces usually enjoyed and Plymouth was recaptured with little trouble.

The 1600 prisoners taken by the Confederates were less important to them than the large quantity of stores, ammunition, and coal they found at Plymouth. Best of all from the Confederate point of view was the recovery

---

2. Daniel Ammen, *The Atlantic Coast* (New York: Charles Scribner's Sons, 1883), p. 203.
3. *NC,* pt. 4, p. 45.

of control of two of North Carolina's richest counties.

Some Confederates wanted the *Albermarle* not to venture to offer battle to the Union gunboats, but to remain at Plymouth as a constant threat to them. Mallory answered this suggestion by saying:

> She was not designed as a floating battery, merely, and while her loss must not be lightly hazarded, the question of when to attack the enemy must be left to the judgment of the naval officer in command [of her], deciding in view of the relation she bears to the defenses of North Carolina.[4]

Her captain decided to use her to support an attempt to recapture New Bern and she left Plymouth for that purpose on May 5, 1864.

She found the *Mattabesett, Miami, Sassacus,* and *Wyalusing,* mounting a total of 58 guns, waiting to greet her in Batchelor's Bay at the mouth of the Roanoke River. The Union plan of attack was for these gunboats, steaming in line ahead, to circle around the *Albermarle* as close aboard as they could without damaging their sidewheels. Each vessel was to fire whenever possible and ram at her captain's choice. The *Miami* was also to use a spar torpedo she carried if an opportunity to do so could be found.

A deadly waltz began, with the *Albermarle* scoring two direct hits on the *Mattabesett* while the latter's 100-pound projectiles bounced harmlessly off of the ironclad's sides.

The *Sassacus* seized an opportunity to ram the *Albermarle* on her starboard side, not really expecting to sink her, but hoping the other gunboats could damage her severely on the other side while she was held comparatively helpless. At the end of a run of 300 or 400 yards the *Sassacus,* carrying 30 pounds of steam (50 percent

4. Ibid., pt. 4, p. 92.

more than normal) struck the *Albermarle* with "a crash that shook the [Union] ship like an earthquake."[5]

With the *Sassacus* pressing hard against the *Albermarle*'s side in an effort to force her under water the Confederate vessel was unable for a few minutes to bring either of her two guns to bear upon her enemy. In that time the *Sassacus* fired three shots from a 100-pound rifle, but she might as well have used blank cartridges for all the damage they did to the ironclad. When the *Albermarle* steamed ahead, causing the two vessels to swing partly alongside of each other, her stern gun was fired at a range of only a few feet. The shot hit the *Sassacus* near the bow, well above the water line, passed through a stateroom, went diagonally across the berth deck, and out of the starboard side without hurting anyone, though it upset a sofa on which an officer was lying. The *Albermarle*'s next shot, from her forward piece, burst through the *Sassacus*'s dispensary bulkhead, a coal bunker, the engine room (where it cut a three-inch iron stanchion), and damaged a boiler before it came to rest in a stateroom. Steam escaping from the boiler "with a shrill scream that drowned for an instant the roar of the guns,"[6] killed one man instantly, fatally scalded four others, injured 10 seriously, and five slightly.

After the *Sassacus* was disabled the other gunboats fought the *Albermarle* until, at nightfall, she steamed back into the Roanoke River battered, but by no means crippled.

About three weeks later (May 25) a party of five volunteers from the *Wyalusing* undertook to sink the *Albermarle* at her mooring in Plymouth. Loading two torpedoes, each containing 100 pounds of gunpowder,

5. Robert U. Johnson and Clarence Buel (eds.), *Battles and Leaders of the Civil War* (New York: The Century Co., 1884–87), 4: 629.
6. Ibid., 4: 630.

into a small boat they rowed up a branch of the Roanoke called the Middle River to a point nearly opposite Plymouth. Landing there, they carried the torpedoes on a stretcher across a swampy island. Coxswain John W. Lloyd and coal heaver Charles W. Baldwin then swam across the Roanoke pulling a line by means of which they hauled the torpedoes to the shore a short distance above Plymouth. While Lloyd stayed where the pair landed, holding the end of a guide rope, Baldwin swam toward the *Albermarle,* intending to place a torpedo on each side of her bow, then to signal to a man stationed on the island to explode them by electricity. When a sentry discovered Baldwin a few yards away from the *Albermarle* he cut the guide ropes and let the torpedoes drift away.

All of the men eventually returned to their ship and, even though their mission was unsuccessful, the captain of the *Mattabesett* said of them,

> I can not too highly commend this party for their courage, zeal, and unwearied exertion in carrying out a project that had for some time been under consideration. The plan [for] executing it was their own, except in some minor details.[7]

On his recommendation the men were awarded medals of honor.

In July, Admiral Lee had the tugboats *Bell, Martin,* and *Hoyt* fitted as torpedo boats and sent them to join the naval force in the North Carolina sounds.

As Lee described them these craft carried torpedoes containing 150 pounds of gunpowder, "intended to explode on impact and . . . placed on a pole or rod projecting not less than 15 feet, and if possible 20 feet beyond the vessel using it."[8]

7. *NC,* pt. 4, p. 62.
8. Ibid., pt. 4, p. 91.

Just how Lee expected these vessels to get within 15 or 20 feet of an ironclad without being blown out of the water is unclear.

As Welles, Fox, and everybody else who gave the matter any thought realized the *Albermarle* could support efforts (all too likely to succeed) to recapture New Bern, Washington, Roanoke Island, and Hatteras Inlet and if she could cross the bar at the inlet she might be used to raid seaports along the Atlantic coast. This being so it was agreed that she would have to be destroyed.

During the early part of the summer of 1864 several plans for accomplishing this aim were considered and rejected. Obviously it would be futile to send a fleet of wooden vessels to Plymouth where they would have to contend with shore batteries as well as with the *Albermarle.* Fox suggested lightening a monitor enough to enable it to cross the bar at Hatteras Inlet. Ericsson said this could not be done. Somebody else proposed the use of one or more of Admiral Lee's improvised torpedo boats. This idea was quickly dropped because anything as big as they were was certain to be discovered long before it could get near enough to the *Albermarle* to accomplish its mission.

Finally it was decided to use either a small rubber boat or a steam launch as a torpedo carrier. Captain S. C. Rowan, who had distinguished himself at Elizabeth City and while in command of the *New Ironsides* at Charleston, was given an opportunity to head the proposed expedition. He respectfully declined the offer because he doubted that the plan would work. After several other names had been considered and rejected for various reasons 21-year-old William B. Cushing was asked if he cared to try to sink the *Albermarle.*

Cushing, who had a strong liking for such risky adventures and had already led several daring commando-like raids, replied: "Deeming the capture or destruction of the rebel ram *Albermarle* feasible, I beg leave to state that I am acquainted with the waters held by her, and am willing to undertake the task."[9]

He preferred to use a steam launch rather than a rubber boat and he went to New York in July to supervise the outfitting of two 30-foot picket boats. Each of them was armed with a 12-pound howitzer and equipped with a torpedo carried on a boom projecting from its bow to be exploded by means of a contraption (such as Rube Goldberg might have invented) which demanded the proper manipulation of every one of several separate lines to avoid having the whole device fail. Cushing remarked later: "It [had] many defects and I would not again attempt its use."[10]

Toward the end of September, while Cushing was enjoying a brief leave of absence at his home in Fredonia, New York, the two boats started south by way of the Delaware and Raritan and the Chesapeake and Delaware Canals. As they were crossing Chesapeake Bay during the first week in October the engine of Launch No. 2 broke down. Her captain, unfamiliar with the locality, anchored in a creek along the Virginia shore of the bay where a small body of Confederate troops found him. He slipped the cable and tried to get away, but the boat ran aground on a sandbar with the result that he and the crew were captured.

Cushing acidly commented:

> This was a great misfortune and I have never understood how so stupid a thing could have happened. I forget the name of the volunteer ensign to whose care [the launch] was intrusted, but am pleased to know he was taken prisoner. I trust that his bed was not of down or his food that of princes while the was in rebel hands.[11]

---

9. Ibid., pt. 4, p. 86.
10. Ibid., pt. 4, p. 119.
11. Ibid., pt. 4, p. 120.

Gustavus V. Fox, Assistant Secretary, U.S. Navy. Courtesy Naval History Division, Navy Department, Washington, D.C.

Admiral David G. Farragut, USN. Courtesy Naval History Division, Navy Department, Washington, D.C.

U.S.S. Hartford. *Courtesy Naval History Division, Navy Department, Washington, D.C.*

Rear Admiral David D. Porter. *Courtesy Library of Congress.*

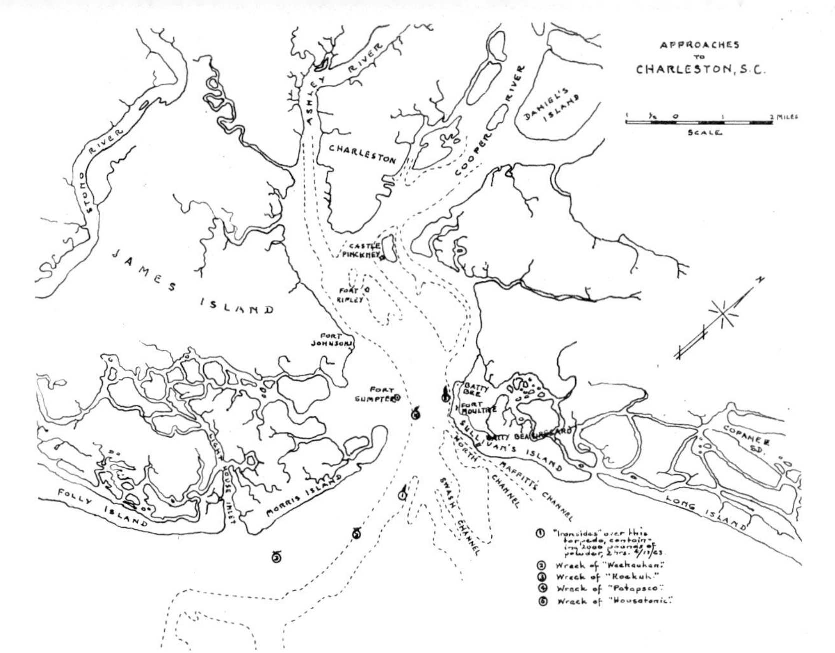

Approaches to Charleston, South Carolina. Courtesy Naval History Division, Navy Department, Washington, D.C.

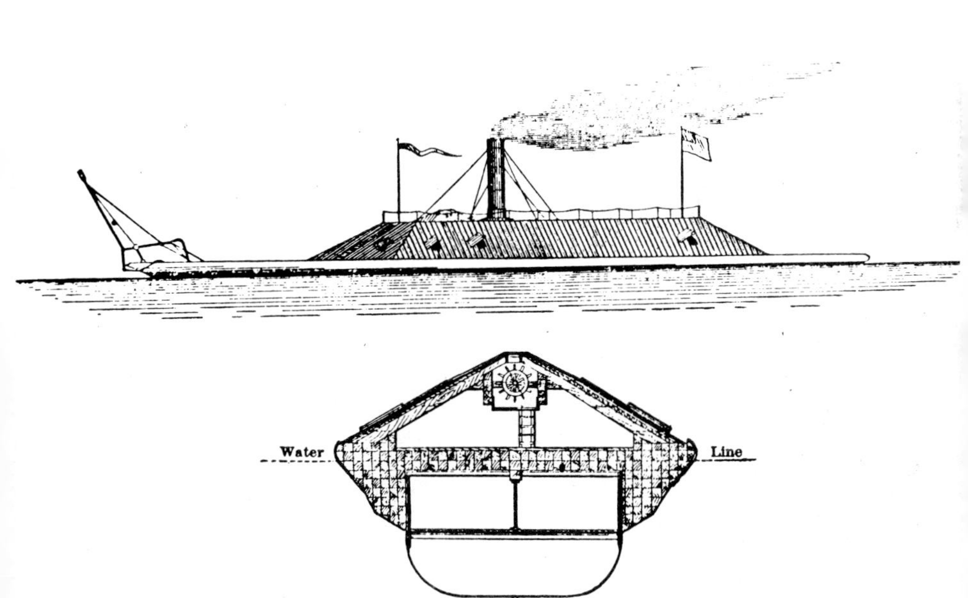

C.S.S. Atlanta. *From* The Atlantic Coast. *Charles Scribner's Sons, 1883.*

C.S.S. Hunley. *Courtesy Naval History Division, Navy Department, Washington, D.C.*

Lieutenant William B. Cushing's launch. Courtesy Naval History Division, Navy Department, Washington, D.C.

C.S.S. Albemarle *attacked and sunk by Lt. Cushing's launch. Courtesy Naval History Division, Navy Department, Washington, D.C.*

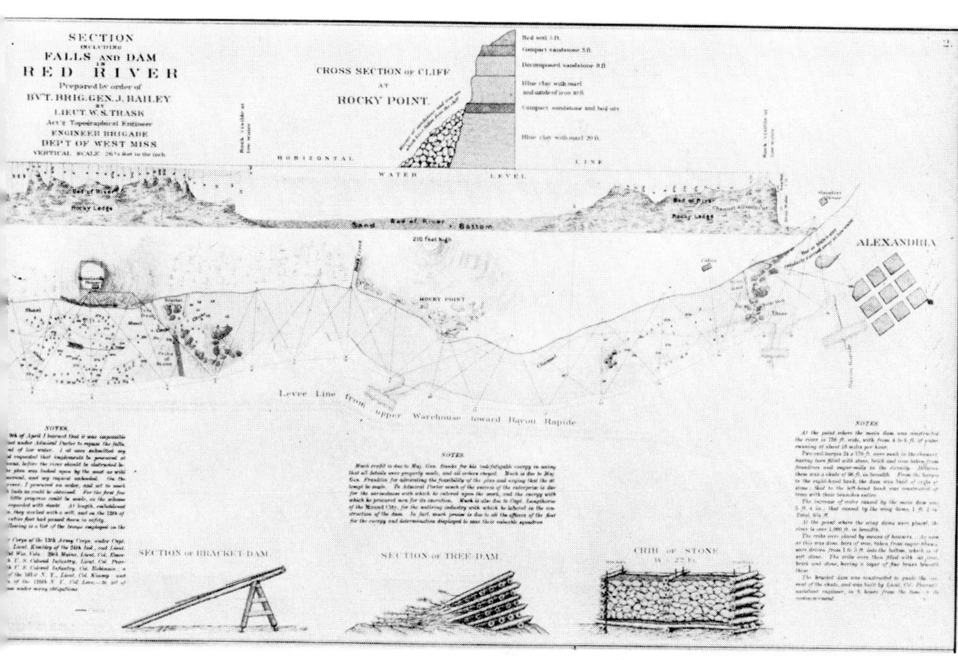

Red River dams. From the War Atlas. *Courtesy of the National Archives, Washington, D.C.*

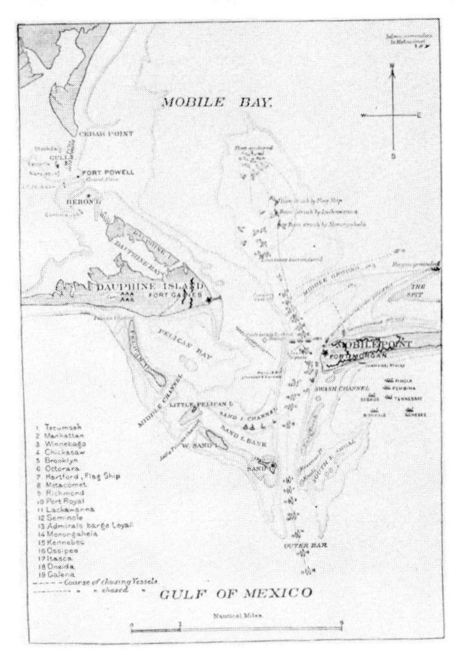

From Battles and Leaders of the Civil War. *The Century Co., 1884-87.*

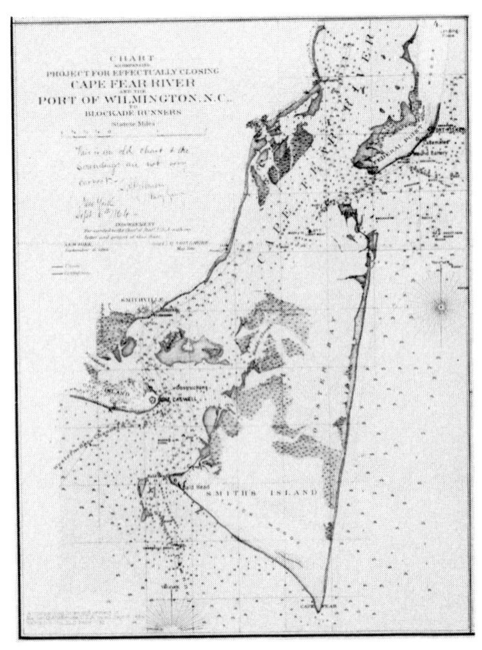

Chart: Cape Fear River approaches to Wilmington, North Carolina, from the War Atlas. *Courtesy of the National Archives, Washington, D.C.*

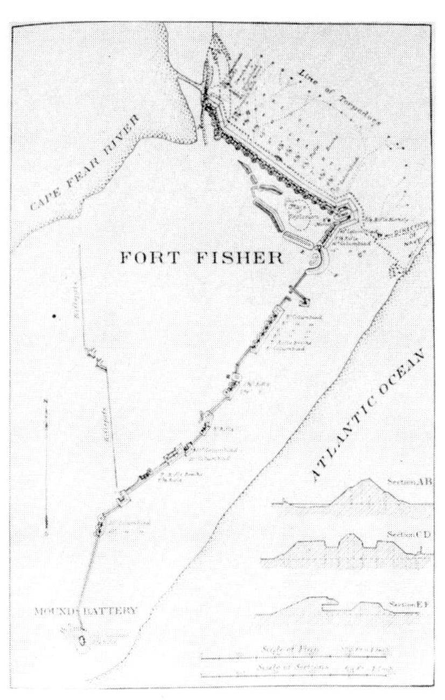

Plan and sections of Fort Fisher. From Battles and Leaders of the Civil War. *The Century Co., 1884-87.*

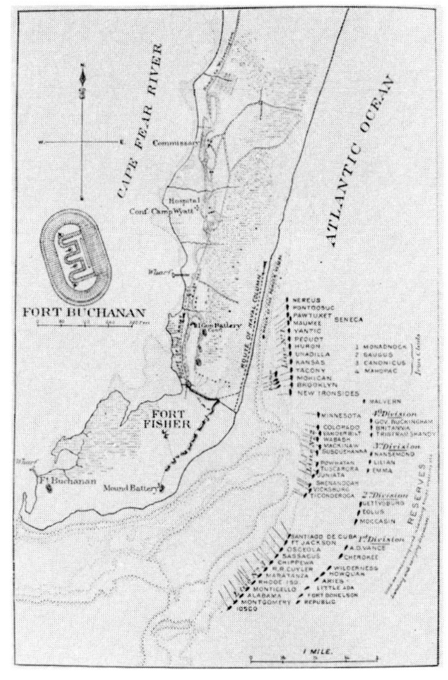

Map of the naval and military attacks on Fort Fisher, January 15, 1865. From Battles and Leaders of the Civil War. *The Century Co., 1884-87.*

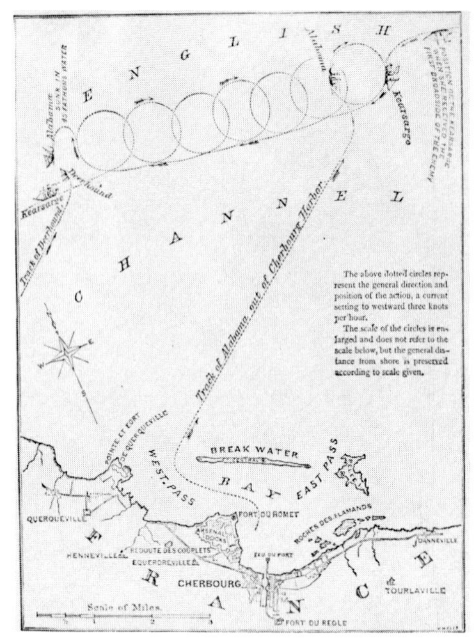

Chart of the battle between U.S.S. Kearsarge *and C.S.S.* Alabama *off Cherbourg, France. From* Battles and Leaders of the Civil War. *The Century Co., 1884-87.*

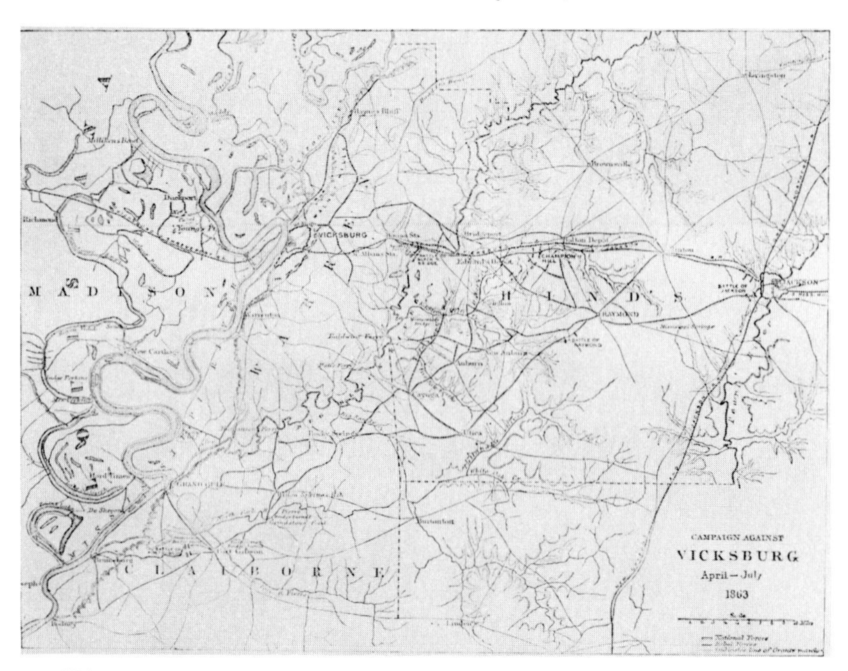

The campaign against Vicksburg, April-July 1863. From Battles and Leaders of the Civil War. *The Century Co., 1884-87.*

Launch No. 1 arrived safely at Hampton Roads where Cushing joined it on October 10. At this time he told his men for the first time what he planned to do and said they were at liberty to go along with him or not as they pleased. All seven of them volunteered to accompany him on the mission.

To obtain some solid information on which Cushing could base his plans Acting Master's Mate John Woodman made three daring daylight reconnaissances of Plymouth. Returning from the first on July 25 he said, "The town appeared very quiet; very few persons were moving about; I could hear the blacksmiths and carpenters at work in the town near the river," and the ram was "lying at the wharf near the steam sawmill." After the second one, made at 10 A.M. August 23, he reported that "the ram *Albermarle* was lying alongside of the wharf, . . . protected with timbers extending completely around her." On October 15 he found the *Albermarle* at her usual berth and learned that the Confederates had stopped trying to raise the *Southfield*.[12]

After spending several days searching for the missing launch Cushing and his men reached Roanoke Island about nightfall October 23. Three days later the launch was towed to the mouth of the Roanoke River eight or 10 miles below Plymouth. Taking her departure at 9 P.M. the launch ran aground before she had traveled more than a couple of miles. At 2 A.M. the following day, soon after she was refloated, she was stopped by a Union picket boat. By the time Cushing had identified himself it was so late the expedition had to be postponed.

This delay may have been providential, for if Union pickets could hear the launch's engine so could Confederates. On returning to the ship where he was making his headquarters Cushing had the carpenter build a box

12. Ibid., pt. 4, pp. 90, 108, 121.

to cover the engine and muffle its noise. Assured that the work would be done in time he planned to try his luck again after dark October 27.

Before Cushing was ready to start a fugitive slave told him there were some Confederate pickets stationed on the hurricane deck of the sunken steamer *Southfield*. In view of this information he decided to have the launch tow a cutter with 13 men in it; if they could not overcome the pickets they could at least create a diversion.

Toward nightfall it began to rain, an ideal turn of events for Cushing's purpose. Leaving the *Otsego* about midnight the launch crept upstream as quietly as possible in the shadows of the trees along the bank. When it passed the *Southfield* without being discovered Cushing began to hope his force (27 men besides himself) could board the *Albermarle,* capture her, and take her down the river to join the Union fleet. However, as the launch and the cutter neared Plymouth a dog barked loudly, alerting a sentry who spotted the boats. A bonfire kindled to light up the river enabled Cushing to see the *Albermarle* tied alongside of a wharf with the pen of logs described by Woodman extending about 30 feet from her sides. (The Confederates had learned of the arrival of the launch at Roanoke Island; its purpose was easily deduced and appropriate precautions had been taken.) Dropping the cutter's towline, Cushing steered the launch through a circle which took her almost to the far side of the river, 150 yards away, then, running at full speed, he headed directly toward the log pen about amidships of the *Albermarle*. The momentum carried the launch a third or more of her length over the log she struck. By this time the Confederate gunfire had become heavy, but, as Cushing remarked: "A dose of canister [from the launch's howitzer] at short range served to moderate their zeal and disturb their aim."[13]

13. Ammen, *The Atlantic Coast,* p. 212.

Even so, several men in the launch were wounded and the back of Cushing's coat was torn by a charge of buckshot—if he had not been standing ramrod straight he probably would have been killed.

As Cushing lowered the torpedo boom the men in the launch found themselves staring into the mouth of a 9-inch cannon, but fortunately for them the explosion of the torpedo preceded the discharge of the gun by a fraction of a second; the concussion lifted the *Albermarle* slightly and 100 pounds of canister flew harmlessly over their heads. The *Albermarle,* with a hole in her side as big as a barn door, sank almost immediately.

At the instant the torpedo exploded Cushing shouted: "Every man for himself," dropped his revolver overboard, stripped off his coat and shoes, and dived into the river. Most of his men followed his example, though a few of them surrendered on the spot. After swimming about half a mile downstream Cushing came upon one of his crew who was almost exhausted. Cushing tried to help the man to reach the shore, but only Cushing made it and he was too weak to crawl out of the water. At daybreak he hid in a swamp. Some hours later he learned from a Negro that the *Albermarle* had been sunk. The following night he procured a skiff in which he made his way to the gunboat *Valley City,* the only one of the party who escaped entirely: two of the others drowned and 25 were captured.

Congress commended Cushing for his bravery and enterprise and he was promoted to the rank of lieutenant commander.

12
Texas and the Red River

After Vicksburg and Port Hudson were finally taken in the summer of 1863 Grant and Farragut earnestly recommended that an early move be made against Mobile. This excellent advice was not taken by their superiors because about the time it was offered France established a puppet empire in Mexico. Acting on the ground that the Mexican debt to France was in arrears, Emperor Napoleon III prevailed upon the Austrian Archduke Maximilian to accept the "throne of Mexico" and sent 50,000 troops to that country to support him. In these circumstances the authorities in Washington considered it more important to move into Texas than against Mobile. Accordingly General Banks was instructed, on July 24, 1863, immediately to prepare part of his Army of the Gulf for an expedition to Texas.

Halleck, now general in chief of the Union Armies, strongly urged Banks and Porter to consider a combined expedition into northern Texas by way of Shreveport, Louisiana, on the Red River, but the choice of a route was left up to Banks.

Because the Red River would be too low to permit

water transportation to Shreveport, 700 miles from New Orleans, until the spring of 1864, Banks decided to send a force along the Gulf coast to Sabine City at Sabine Pass where the Sabine River, which separates Louisiana from Texas, flows into the Gulf. He thought that if his troops could land there they could reach Houston, the railway center of Texas, less than 100 miles away, before the Confederates could effectively oppose them.

(Three of Farragut's vessels had taken Sabine City late in September 1862, but General Butler, then the commanding officer of the Army of the Gulf, had not been able to spare any troops to garrison the place so it had not been held.)

Banks did not have enough transports to carry all of the troops he planned to use against Sabine Pass at one time so the first contingent, a division of 4000 officers and men, departed from New Orleans on September 5, 1863, in a convoy protected by the third and fourth class gunboats *Arizona, Clifton, Sachem,* and *Granite City,* the best vessels of sufficiently shallow draught attached to the West Gulf Squadron.

The naval commander of this expedition and Banks intended to have the gunboats, with about 200 Army sharpshooters in them, surprise the Confederates and drive them from their works before the troops attempted to land. Instead of adhering to this excellent plan the convoy arrived at Sabine Pass late Sunday, September 6, and spent all of Monday in full view of the shore. The Union vessels finally crossed the bar at 8 A.M. Tuesday, then, for some unexplained reason, they anchored until 3:30 P.M. At that time the *Sachem* and *Arizona* started up the Louisiana, or right-hand channel, while the *Clifton* and *Granite City* entered the Texas, or left-hand channel. Wisely waiting for them to come to within point blank range, the Confederates finally opened a devastating fire on them. The *Sachem* was almost immediately disabled by a shot through her boilers and was surrendered after her magazine had been flooded. Another shot cut the *Clifton*'s wheel ropes and she drifted ashore under the Confederate guns. She fought until 10 men had been killed and nine seriously wounded; then her captain lowered her colors to stop the slaughter. With half of its naval strength gone, the expedition returned to New Orleans without even attempting to land any of the troops.

Late in October 1863 Banks personally led some troops to Brazos Island, Texas, at the mouth of the Rio Grande. He left New Orleans on the 26th with enough transports for 3500 men convoyed by the *Monongahela, Owasco,* and *Virginia.* As happened so many times during the Civil War this expedition enjoyed several days of fine weather followed by a severe storm. A Texas norther, which began during the afternoon of October 30, almost wrecked the transport *Zephyr* and did cause the steamer *Union* to sink while she was towing the S.S. *Empire City*. Fortunately boats from the two ships saved all hands in both of them.

On November 2, the day after the expedition reached Brazos Island, the bar was buoyed and the troops landed without opposition. However, the force Banks had with him was too small to exploit the foothold it gained and Halleck was unwilling to reinforce it, partly because it was so far away from New Orleans, partly because he still preferred the Red River route into Texas.

Even now Halleck did not specifically order Banks to send an expedition up the Red River, but as an inducement for him to do so he was promised the help of a substantial number of troops from General William T. Sherman's Army of the Tennessee and much of Admiral Porter's Mississippi Squadron. In these circumstances Banks unenthusiastically prepared to follow Halleck's plan.

The expedition could not start until the Red River began its annual rise, an event which usually occurred in March, and the movement would have to be made quickly because Sherman would need his troops back in time to take part in the spring campaign east of the Mississippi. With these facts in mind Banks and Sherman agreed at a conference they held in New Orleans on March 1, 1864, that Sherman would send 10,000 men to meet Porter at the mouth of the Red River in time to reach Alexandria, Louisiana, on March 17; that Banks, marching from Franklin, Louisiana, should be at Alexandria on the same date; and that the movement should be conducted rapidly enough to permit Sherman's troops to be back to the Mississippi 30 days after they entered the Red River. Tacitly they relied on Porter's gunboats to spearhead their movement and to keep the expedition's supply lines open.

After their conference ended Banks invited Sherman to remain in New Orleans until March 4 to witness the installation of the recently elected pro-Union government of Louisiana. Banks had arranged for this occasion such ceremonies as the playing of the *Anvil Chorus* by the massed bands of the Army of the Gulf, the firing of cannon by electricity, etc., and he thought Sherman might like to witness these remarkable events. If by any chance Sherman found this invitation attractive he had the will power to resist it. By March 6 he was back in Vicksburg where he immediately issued orders to General Andrew J. Smith who was to command the troops from the Army of the Tennessee assigned to the Red River expedition.

On the last day of February 1864 Porter ordered the stern-wheel monitor *Osage,* the ironclad gunboats *Lexington* and *Conestoga,* and the tinclads *Fort Hindman* and *Cricket* to reconnoiter the Black and Ouachita Rivers. These boats met with no resistance until they neared Trinity, Louisiana, on March 1, when some Confederate sharpshooters fired on them. A brisk return fire of grapeshot and canister drove off the attackers and the boats made their way above the city before they anchored for the night. They had a more serious encounter with some Confederate field artillery the following day. The leading boat, the *Fort Hindman,* had her starboard engine disabled by one of the 27 hits she sustained. After going as far as Catahoula Shoals and Bayou Louis the boats turned back because the water was falling fast. They reached the Mississippi again on March 3.

Smith's transports joined Porter who was waiting at the mouth of the Red River with the ironclads *Eastport, Lexington, Essex, Benton, Choctaw, Chillicothe, Ozark, Louisville, Carondolet, Pittsburg, Mound City, Osage,* and *Neosho,* the wooden steamers *Lafayette* and *Ouachita,* the tinclads *Fort Hindman, Gazelle,* and *Cricket,* and the headquarters boat *Black Hawk* on March 11.

Early the following morning these boats and Smith's transports entered the Red River, which they found unusually low for this time of the year. (It was low partly because the Confederates had diverted some of its headwaters, but this fact was unknown to Porter and Smith.)

As previously arranged by Porter and Smith the *Eastport, Essex, Ozark, Osage, Neosho, Lafayette, Choctaw, Fort Hindman,* and *Cricket* pushed ahead to remove some obstacles in the river below Fort De Russy, the expedition's first objective. At the same time Porter went with the *Benton, Pittsburg, Louisville, Mound City, Carondolet, Chillicothe, Ouachita, Lexington,* and *Gazelle* into the Atchafalaya to cover the landing of the troops at Simmsport from which place they were to march to the fort.

A newspaper correspondent who accompanied the expedition wrote at this time:

> I should not be a faithful historian if I omitted to mention that the conduct of the troops . . . is becoming very prejudicial to our good name and to their efficiency. A spirit of destruction and wanton ferocity seems to have seized upon many of them, which is quite incredible. At Red River Landing they robbed a house of several thousand dollars in specie, and then fired the house to conceal their crime. At Simmsport, a party of them stole out, and robbed and insulted a family two miles distant. In fact, unless checked by a summary example, there is a danger of our whole noble army degenerating into a band of cut-throats and robbers.[1]

During the afternoon of March 14 the boats heading directly up the Red River, slowed by the difficulty of navigating the *Lafayette* and *Choctaw* in the narrow, winding stream, came upon a raft of logs well secured to either bank and two rows of piles driven into the bed of the channel with a mass of driftwood heaped upon them some distance below Fort De Russy. By the time a passage had been forced through these obstructions the troops had taken the fort without the help of the gunboats except for a few shots they fired at long range.

Porter left the *Benton* and *Essex* at Fort De Russy to protect an army detachment engaged in destroying the works, sent the *Eastport, Lexington,* and *Ouachita* ahead to try to overtake some Confederate boats, and followed with the rest of his fleet convoying the transports. Most of the Confederate boats got away; one of them, the S.S. *Countess,* ran aground and was set on fire by her own crew.

Two days after the surrender of Fort De Russy the gunboats and transports reached Alexandria where they anchored below the Falls of Alexandria, a two-mile-long

1. F. Moore, *Rebellion Record* (New York: G. P. Putnam; D. Van Nostrand, 1861–67), vol. 8, docs., p. 430.

rapids above the town. The river was so low that Porter was afraid any heavy boats sent beyond the falls would be stranded there. He suggested to Banks, who had arrived at the rendezvous a week behind schedule, that his troops should march to Shreveport, 340 miles upstream, without naval support. Banks was so reluctant to do this that Porter agreed to send the *Eastport, Osage, Mound City, Carondolet, Pittsburg, Louisville, Chillicothe, Ozark, Neosho, Lexington, Fort Hindman,* and *Cricket* to Shreveport.

The transports made their way above the falls fairly easily and most of the gunboats got over them without too much trouble, but it took two and a half days to haul the *Eastport* past them by main force and the Army hospital boat *Woodford* struck a rock and sank.

With Banks's troops moving overland and Smith's in their transports, convoyed by the gunboats and tinclads, the expedition reached and occupied Grand Ecore, Louisiana, without opposition on April 3.

Three days later Banks, Smith, and Porter left Grand Ecore planning to meet again at Springfield Landing, Louisiana, 110 miles farther up the river.

The route Banks followed led through some densely wooded country traversed by a single road which in many places was too narrow for two wagons to pass each other. His army reached Pleasant Hill, Louisiana, at nightfall April 7, with its van (a cavalry division) more than 20 miles ahead of its rear guard.

The next day the cavalry, which had been reinforced with a brigade of infantry, was engaged by a substantially superior force about 15 miles beyond Pleasant Hill. After being pushed back for some distance the advance corps gave way and fell back on a train of 150 wagons and 18 or 20 pieces of artillery immediately in its rear. Because it was impossible to turn the wagons around in the narrow road their drivers were thrown into utter confusion and what had been a disorderly retreat quickly became a rout.

At 4 P.M. General William B. Franklin came upon the scene, saw what was happening, and sent word to General William H. Emory to form a line of battle at Pleasant Grove, two or three miles to the rear. Emory, moving rapidly in accordance with his orders, met a disorganized crowd of cavalrymen, infantrymen, wagons, ambulances, and loose animals through which his division had to force its way, using violence to do so. He deployed his troops and they checked the Confederates after about 15 minutes of hot fighting.

During the night Banks fell back to Pleasant Hill where the Confederates attacked him at 5 P.M. the following day. They were repulsed so decisively that most of the Union troops and officers thought they had won a victory and the Confederates admitted having suffered a setback. However, Banks (who had, with good reason, come to be known as "Nothing Positive" Banks) decided for reasons best known to himself, to retire to Grand Ecore, where he arrived on April 16.

Meanwhile, Porter and Smith, unaware of Banks's misfortune, pushed up the river with the gunboats and transports. Although the river was narrow, crooked, and full of snags, they reached Springfield Landing during the afternoon of April 10, right on schedule. Here they found the river blocked by a steamboat with its bow touching one bank, its stern the other, and its broken hull resting on the bottom. They were still wondering what to do about this obstacle when a courier rode in with news of Banks's retreat and orders for Smith's troops to return to the Mississippi immediately.

The gunboats, of course, turned back with the transports to protect them as they made their way downstream. On their way back the boats were frequently harassed by Confederate troops, heavily concentrated along the right bank, or southern side of the river. At places where the banks were high the Confederates could fire almost with impunity on the boats whose guns could not be elevated enough to reply effectively.

Late one afternoon the transport *Alice Vivian* and the gunboat *Osage* were aground in midstream with three transports trying to free them, the transport *Hastings* was tied up at the right bank having a damaged paddle-wheel repaired, the gunboat *Lexington* was afloat near the northern shore, and the transport *Rob Roy,* with four siege guns on her deck, had just appeared from upstream when 2000 Confederate infantrymen and a battery of four fieldpieces attacked them. The *Hastings* quickly left her dangerous mooring while the gunboats and the *Rob Roy*'s siege guns opened fire. Two hours later the Confederates fell back, having lost 700 men, including their commanding officer, General Thomas Green of Texas.

Despite a number of encounters similar to the one just described and a variety of navigational hazards all of the boats that had been to Springfield Landing got safely back to Grand Ecore on April 15.

Porter, who was uncertain how far and how fast Banks intended to retreat, waited for the General at Grand Ecore with four shallow draught gunboats while the rest of the fleet was sent to safety, as Porter supposed, at Alexandria.

When Porter learned that the *Eastport* had been sunk by a torpedo eight miles below Grand Ecore on the day she left that place he immediately went to her aid. A few days later he found out that Banks had decided to fall back to Alexandria as quickly as he could without regard to the situation in which this would leave the *Eastport.* Because the boat was too valuable to be abandoned without at least an effort to save her Porter stood by her with a couple of tinclads and two salvage boats equipped with powerful pumps.

With all hands working around the clock it took six days to raise the *Eastport*. After she was refloated she made a run of 20 miles, then grounded on a sandbar where she spent the next 24 hours. A few miles farther downstream she went aground for another day. Finally, five days after she was first raised, she became hopelessly stuck on some snags near Montgomery, Louisiana, 50 miles from the place where she had been torpedoed and 60 miles from Alexandria. Her people were transferred to another boat, eight barrels of gunpowder were placed in her forward casemate, the same number in the after one, several in the engine room, and she was blown up.

Just as preparations for the destruction of the *Eastport* were completed 1200 Confederates made a rush for the *Cricket,* tied up at the left bank with Porter on board. This attack was quickly repelled, but the next one had a different outcome.

A short distance above the mouth of the Cane River a number of Confederate fieldpieces opened fire upon the five boats that had been trying to salvage the *Eastport*. (Porter said there were 18 guns; a Confederate report put the number at four.)[2] The *Cricket,* in the lead, stopped to help to cover the other boats, but Porter ordered the pilot to ring for full speed ahead. As the boat steamed past the battery more than half of her crew of 50 men were killed or wounded. The salvage boat *Champion No. 5,* immediately behind the *Cricket,* had her boiler hit by a 12-pound shot and nearly 200 of those on board, most of whom were escaped slaves, were scalded to death. (This was the only shot fired during

2. A. T. Mahan, *The Gulf and Inland Waters* (New York: Charles Scribner's Sons, 1883), p. 201, ftn.

the Civil War that caused more casualties in a naval vessel than did the one that hit the *Mound City* in the White River.) The *Cricket* fought the battery from below for a little while, then went downstream to send help back to the other boats.

At noon the following day, when no help had reached them, two of the boats ran past the battery and the third was captured.

In the meantime the other gunboats had made their way down to the Falls of Alexandria, only to be stopped there because there was less than four feet of water and they needed at least seven feet. Ten gunboats and two towboats were thus trapped at a place that Banks's army might abandon at any time.

A volunteer naval officer, who had been a civil engineer for 14 years before the war, said he believed he could build a cofferdam around the falls and blast a channel through them. This idea was utterly ignored by Porter, to whose attention it was brought.

As the situation grew worse a more desperate remedy was suggested by Lieutenant Colonel Joseph Bailey of the Fourth Wisconsin Volunteers. He proposed to dam the river below the falls to back up enough water to float the boats over them. He believed his plan would work because he had done the same sort of thing in similar, though peaceful, circumstances on smaller streams in the Northwest. Porter, who, playing on words, said, "If damning would get the fleet over it would have been done long ago,"[3] had almost no faith in Bailey's plan. However, there seemed to be no other hope of saving the boats so Bailey was allowed to begin work on May 1. Several hundred wagons, about 1000 draft animals, and 3000 men—including a regiment composed largely of lumberjacks from Maine—were put at his disposal by

3. Moore, *Rebellion Record,* vol. 11, docs., p. 11.

Banks. Although the Navy was to be benefited by Bailey's efforts, the only naval officer who seemed to have any hope that his plan would work and substantially helped him to execute it was the volunteer lieutenant commanding the *Mound City*.

At the place where the dam was to be built the Red River was nearly 800 feet wide and had a current of nine or 10 miles an hour. Partly to save time, partly because different kinds of building material were available on the different sides of the stream, work was carried on simultaneously from both banks. A so-called tree dam, formed of untrimmed trees fastened together with heavy pieces of timber and laid with their branches pointing upstream to hold bricks dumped among them and to catch debris, was built from the left bank. A crib dam, composed of wooden structures resembling corn cribs, filled with bricks obtained by tearing down a local sugar mill and stones found nearby, extended from the right bank where there were fewer trees. A space 150 feet wide between the ends of the wing dams was closed by sinking four barges loaded with bricks across it.

During the hot days and chilly nights the dam was under construction trees centuries old were cut down and dragged by teams of horses, mules, or oxen to the river's banks where they were pulled into the final positions; flatboats filled with stone or bricks, guided by ropes held by men on shore, were floated down by the current while other boats were hauled upstream; hundreds of men loaded boats, pushed wheelbarrows over catwalks, and contributed in various ways to the work. The scene, striking enough in daylight, was even more remarkable at night with half visible figures moving to and fro and huge bonfires burning on both sides of the river.

Officers and men of the Union Army and Navy, residents of Alexandria, camp followers, and others watched the progress of the work and discussed the chances of its success—the consensus was that it would fail. The gloomy prophets among the Union men, the hopeful ones among the Confederate sympathizers, both proved to be wrong. By May 8 enough water had been backed up to permit three shallow draught boats to cross the falls and anchor just above the dam. Meanwhile everything possible was done to lighten the other boats. Their ammunition, chains, anchors, most of their coal, etc., was carried around the falls in wagons, the armor was stripped from the ironclads, and because it was too heavy to be carted downstream to be reinstalled it was sunk in 30 feet of water at a spot where shifting sand would quickly cover it.

As the dam neared completion after eight days of Herculean labor it began to seem possible that the Mississippi Squadron might be saved. Then, about 5 A.M. May 9, two of the barges which formed part of the structure were swept aside by the pressure of the water backed up behind them. However, they swung into such a position that they formed a chute over the Falls. Porter immediately ordered the *Lexington,* the only boat with steam up, to run through the gap. Pitching like a canoe shooting a rapids, she tore downstream, hung for a moment on some rocks, then, with a harsh grating sound, ran into deep water. The *Neosho,* quickly following the *Lexington,* almost came to grief because her pilot lost his nerve and rang the engine down when it should have been kept at full speed ahead. However, the *Neosho*'s hatches had been battened down and, though her hull was completely under water for several seconds, the current carried her to safety. The *Fort Hindman* and *Osage* also got through before the water dropped too far for any more boats to make it.

Undaunted by the partial collapse of their dam, Bailey

and his men set to work to repair it. They succeeded so well that within four days the rest of the fleet was able to cross the falls. A week later the expedition was back in the Mississippi with nothing to show for the work it had done, the risks it had taken, and the casualties it had suffered.

13
Mobile Bay

As already mentioned the authorities in Washington expected Farragut, with the aid of General Butler's Army of the Gulf and Captain Davis's Western Flotilla, to gain control of the Mississippi River during the summer of 1862 after which Farragut and Butler were to turn their attention to Mobile. Because this timetable was upset by the stubborn defense of Vicksburg and the conduct of the abortive Red River expedition Mobile remained undisturbed for a long time.

Toward the end of January 1864 Farragut resumed command of the West Gulf Squadron after five months spent in the North while the *Hartford* was being repaired. At this time he began to make plans, which he carried out six months later, for an attack on the defenses of Mobile Bay. His primary objective was not to take the city of Mobile, situated at the head of the bay about 30 miles from the Gulf of Mexico; its capture could wait. If he could gain control of the lower part of the bay and its rather narrow entrance the only important Gulf port still in the hands of the Confederates would be cut off from commerce with the world outside of the Confederate States. (His planning so far ahead, so wisely, and so carefully are all characteristic of the man.)

The tactical situation at Mobile Bay was somewhat like the one at New Orleans in that it would be necessary for an attacking force to pass two forts and contend with a small squadron of wooden and ironclad vessels.

One of the Confederate works, Fort Morgan, on Mobile Point to the east of the Main Ship Channel, mounted twenty-two 32-pound smoothbores, four 10-inch columbiads, three 8-inch rifles, four 5.82-inch rifles, and two 32-pound rifles. The other work, Fort Gaines, on Dauphin Island about three and a half miles to the west of the channel, was armed with three 10-inch columbiads, four 32-pound rifles, and 20 smoothbores of various calibers.

The entrance to Mobile Bay was too wide to be closed by means of a boom such as the one that had been used below New Orleans, but a double row of stakes, intended to prevent the passage of shallow-draught vessels, extended from Dauphin Island to the western edge of the Main Ship Channel which ran close aboard Fort Morgan; and the channel was heavily mined except for an opening 100 yards wide immediately off Mobile Point.

The Confederate squadron consisted of three wooden gunboats—the *Morgan,* two 7-inch rifles and four 32-pounders; the *Selma,* a 6-inch rifle, two 9-inch and one 7-inch shell guns; the *Gaines,* a 7-inch rifle and five 32-pounders—and one ironclad, the *Tennessee.*

The *Tennessee,* the most powerful battleship built from the keel up in the Confederate States, measured 209 feet long, 48 feet in beam, and drew about 14 feet fully loaded. Her casemate, or shield as the Confederates called it, was framed with closely spaced yellow pine timbers 13 inches square and sheathed with yellow pine planking 5½ inches thick laid horizontally and oak planking 4 inches thick laid vertically. Her laminated armor, fastened with 1¼ inch bolts, varied from 6 inches thick forward to 5 inches along the sides and across the after end of the casemate. To protect the ship against damage by ramming the armor extended a couple of feet below the water line at the same angle as the casemate's sides, then was reversed and brought back to meet the hull. Thus she had a sharp-edged sponson about 8 feet wide on each side. These sponsons were carried around the bow to form a ram that would not break off as the *Virginia*'s had done. With only two 7-inch and four 6-inch rifles the *Tennessee* was not heavily armed for a vessel of her size, but this was the best ordnance her builders had available. Because her non-condensing engines, taken from a river steamboat, were not powerful enough for a heavily built warship the *Tennessee*'s best dependable speed was six knots; she could be pushed to eight knots with some difficulty. Her designer made a bad mistake in having her steering chains run to the rudder head through the end of the casemate in troughs along the deck covered with boiler plate only an inch thick.

The *Tennessee*'s keel was laid in the spring of 1863 at Selma, Alabama, on the Alabama River about 150 miles above Mobile and she was launched the following winter.

While the *Tennessee*'s engines and armament were being installed at Mobile a set of camels was prepared to enable her to cross the Dog River bar in Mobile Bay where there was only nine feet of water. (Camels were pontoons with chains between them. They were sunk on either side of a vessel, which was supported on the chains and lifted when the camels were pumped dry.)

With unquenchable optimism the Confederates expected the *Tennessee* at least to raise the blockade of Mobile and probably to recapture Pensacola and New

Orleans. However in mid-May 1864, while she was making a trial trip in Mobile Bay on a day when the sea was only moderately rough, she shipped so much water her fires were nearly extinguished. Obviously unfit to venture into the Gulf of Mexico, much less to go into the open sea, she was anchored near Fort Morgan to help to defend Mobile Bay.

Farragut, who never came to like ironclads, but had learned to respect them, wanted to obtain some monitors to enable him to cope with the *Tennessee*. In mid-January, 1864 he wrote to Admiral Porter:

> I am . . . anxious to know if your monitors, or at least two of them, are not completed and ready for service; and if so, can you spare them to assist us? If I had them, I should not hesitate to become the assailant instead of awaiting the attack. I must have ironclads enough to lie in the bay and hold the gunboats and rams in check in the shoal water.[1]

As Farragut learned more about the *Tennessee* and heard that some other ironclads were being built at Mobile he suddenly came down with as severe a case of "ram fever" as anyone in the Union Navy ever had. Early in May (while the Confederates were still trying to get the *Tennessee* over the Dog River bar into Mobile Bay) he wrote to Welles:

> I am in hourly expectation of being attacked by an almost equal number of vessels, ironclads against wooden vessels, and a most unequal contest it will be, as the *Tennessee* is represented to be impervious to all their experiments. . . .[2]

He soon recovered his nerve and said to one of his subordinates,

1. *Civil War Naval Chronology* (Washington, D. C.: Naval History Division, Office of the Chief of Naval Operations, Navy Department, n. d.), pt. 4, p. 10. (Hereafter cited as *NC*.)
2. Ibid., pt. 4, pp. 58–59.

> I can see his [the Confederates'] boats very industriously laying down torpedoes, so I judge he is quite as much afraid of our going in as we are of his coming out; but I have come to the conclusion to fight the devil with fire, and therefore shall attach a torpedo to the bow of each ship, and see how it will work on the rebels—if they can stand blowing up any better than we can.[3]

For some reason he did not equip his ships with torpedoes, but he did have the *Brooklyn* fitted with a device for clearing them away.

After receiving a report from Farragut that he had seen the *Tennessee* in lower Mobile Bay, Welles decided to send some monitors to the Gulf. He ordered the *Manhattan,* then being fitted out in New York, and the *Tecumseh,* then assigned to the James River Squadron, to join Farragut's Gulf Squadron, and told Porter to send Farragut a couple of river monitors. Porter protested that his vessels were not suitable for Farragut's use, but sent him the *Winnebago* and *Chickasaw,* both double-turreted vessels.

On July 25, while Farragut was waiting for these craft to arrive, he had some small boats reconnoiter the Main Ship Channel in an attempt to find out what kind of torpedoes, and how many there were, off Fort Morgan; if possible the boats were to cut loose any torpedoes they found. During the next few days another boat and a tug sounded the outer channel and marked the edges of the Confederate torpedo field with buoys.

Up to this time Farragut had paid frequent visits to Mobile Bay, but he had left a subordinate in immediate command of the vessels on blockade there. During the last week in July he took personal charge of the preparations for the attempt to enter the bay. When everything was finally ready he waited for an early morning flood tide to coincide with a westerly wind. He did this because

3. Ibid., pt. 4, p. 63.

he thought that on an incoming tide the torpedoes would be less dangerous since their tops would point away from the ships. He was determined to get his fleet into the bay at any cost and on a flood tide a disabled vessel would drift in the right direction. A westerly wind would blow the smoke from the ships' guns toward Fort Morgan, thus making things difficult for the Confederate gunners.

When General Robert S. Granger arrived at Mobile Bay on August 1 with 2400 men—all that could be spared by the Army of the Gulf—he and Farragut decided they could do the most good by landing on Dauphin Island and investing Fort Gaines.

At 3 A.M., August 5 Farragut checked the weather, found the wind blowing from the southwest, and decided to try to enter the bay that morning. During the next hour and a half the gunboats and battleships were fastened together as they had been at Port Hudson. Since Fort Gaines would be almost out of range on the port side the smaller vessels were fastened on that side of the larger ones to form the following pairs: *Brooklyn-Octorora, Hartford-Metacomet, Richmond-Port Royal, Lackawanna-Seminole, Monongahela-Kennebec, Ossippee-Itasca,* and *Oneida-Galena.* The monitors made up a separate starboard column and were charged with the twin duties of keeping down Fort Morgan's fire and attacking the *Tennessee* as soon as they passed the fort.

The wooden ships were prepared, as they had been at New Orleans, to make them as nearly impregnable as possible.

At first Farragut planned to have the flagship *Hartford* and her consort head the column of wooden vessels because he felt strongly that a commanding officer's proper place was in the van. However, because the *Brooklyn* had four bow guns instead of the usual two and was equipped for clearing torpedoes from her path he accepted the suggestion of his subordinates that she and the *Octorora* take the lead.

When the signal to get under way was made at 5:30 A.M. the fleet steamed slowly ahead to allow the vessels to take their assigned stations while their crews cleared them for action and went to quarters. Forty minutes later the *Hartford* and *Metacomet* crossed the bar and the line of battle was quickly formed. The leading monitor, the *Tecumseh,* opened the game with two shots fired at 6:45 A.M. The fort replied to this challenge at five minutes past seven and the Confederate gunboats stationed themselves in a row running east and west just inside of the torpedo field where they could direct a raking fire on the Union ships, which would have to steer north by east to stay in the channel until they passed Fort Morgan.

As the fleet moved toward the bay the *Tecumseh* started to pass at full speed between two buoys 150 yards offshore abreast of the fort. Just as she reached them a torpedo exploded under her turret with such force that men in the nearby *Manhattan* half thought their vessel had been hit. The *Tecumseh* lurched clumsily a couple of times, then went down by the head like a whale sounding, with her propeller still spinning as she disappeared in 60 feet of water less than 30 seconds after the explosion occurred.

Ninety-three of the *Tecumseh*'s complement of 114 officers and men were lost. The only ones who escaped from her were the pilot and some who had been in the turret. The captain died because he allowed the pilot to precede him from the pilothouse and there was not time enough for both of them to leave it.

A few minutes after the *Tecumseh* sank the *Brooklyn* suddenly came upon a row of buoys which the captain thought might mark the location of some torpedoes. Instead of attempting to clear them away, as his ship was supposedly equipped to do, he rang the engine down, then ordered the *Brooklyn*'s engine and the *Octorora*'s reversed. Their sudden stop threw the rest of the port column into such confusion that the *Hartford* and *Metacomet,* the former steaming at full speed ahead, the latter with her side-wheels turning at full reverse, barely avoided colliding with the *Brooklyn.* Farragut, standing in the *Hartford*'s mizzen rigging, asked the pilot in the top just above him: "What's the matter with the *Brooklyn*? She must have plenty of water there," and ordered the *Hartford* to take her place at the head of the line where it seemed to him she belonged. As the *Hartford* came abreast of the *Brooklyn,* Farragut was told there were torpedoes in the channel. At this he shouted: "Damn the torpedoes! Full speed ahead!"[4] (One of Farragut's subordinates remembered his exact words as having been: "Damn the torpedoes! Full speed ahead, Drayton [the *Hartford*'s captain]. Hard astarboard; ring four bells! Eight bells! Sixteen bells!"[5] If the last two orders really were given they are highly imaginative, since four bells called for maximum speed.)

This famous episode was first publicized by Farragut's son Loyall in his *Life of David Glasgow Farragut* written 15 years after the event and it has been asked if an oral order could have been heard over the din of battle. The answer to this question is that it could have been heard through the speaking tube Farragut had caused to be rigged at the time he undertook to pass Port Hudson months earlier.

The *Hartford* and *Metacomet* surged past the *Brooklyn* and *Octorora* which were taking a severe raking fire from Fort Morgan and backing toward the *Richmond* and *Port Royal.* The *Richmond*'s captain, seeing the

4. Loyall Farragut, *The Life of David Glasgow Farragut* (New York: D. Appleton and Company, 1879), pp. 416–17.
5. *NC,* pt. 4, p. 95.

danger of a collision, ordered his ship and her consort to back their engines lest the four vessels become entangled where, if they should be sunk they would block the channel. After milling around for a few minutes the ships straightened out their line and followed the *Hartford* and *Metacomet* through the buoys near where the *Tecumseh* had sunk. As they did so their crews thought they heard torpedoes knocking against their hulls; some men below decks even believed they heard primers snapping. Fortunately for the fleet there were no more explosions.

The Union ships' heavy, rapid fire enabled most of them to pass Fort Morgan without being badly damaged. However, when the larger vessels drew out of range the Confederates returned to their guns in time to hit the *Oneida* with a couple of rifle shells. One of them burst in the cabin, cut the wheel ropes, and disabled two guns. The other penetrated the improvised chain armor and hit the starboard boiler, disabling the ship and scalding most of the men on watch. However, she was towed past the fort into the bay by the *Galena.*

As the *Hartford* and *Metacomet* advanced over the line of torpedoes they came under a raking fire from ahead and to starboard from the three smallest Confederate gunboats. Fighting mainly with their stern guns at ranges of from 700 to 1000 yards these craft punished the *Hartford* severely. A shell one of them fired burst under a bow gun and disabled it. Another shell killed 10 men and wounded five more of the ship's forward division.

When the *Hartford* and *Metacomet* passed Fort Morgan a few minutes after 8 A.M. the *Tennessee* went into action. She fired first at the *Hartford,* aiming below the water line. The shot struck the water, ricocheted like a skipping stone, and hit the ship so high it did no great damage. Next the *Tennessee* tried to ram the *Hartford,* only to have the faster Union vessel easily evade the blow. The *Tennessee,* trying hard to sink the *Hartford,* followed her for a mile or two, until she found pursuit useless. Then she stood for the *Brooklyn* as if to ram her, but sheered off before they came together and went down the line of Union ships. As they passed close aboard each other the *Tennessee* put two shots into the *Brooklyn* whose broadside left the ram entirely undamaged. An exchange of broadsides between the *Richmond* and *Tennessee* at short range did no harm to either of them. After passing the *Lackawanna* the *Tennessee* suddenly swung around as though she were looking for a vessel to ram. The *Monongahela,* which was equipped with a ram of sorts, countered this maneuver by trying to strike the *Tennessee* at a right angle. With the *Kennebec* still lashed alongside of her the *Monongahela* could not turn fast enough to accomplish her purpose and she gave the ram only a slight, glancing blow on the port side. The *Tennessee* was pushed around and passed close along the port side of the *Kennebec,* damaging her bow planking and carrying away a small boat with its iron davits as a souvenir of the encounter. As the pair rasped by each other a shell fired by the *Tennessee* exploded on the *Kennebec*'s berth deck, seriously wounding an officer and four men. The *Ossippee,* on the *Monongahela*'s quarter when she hit the *Tennessee,* also tried to ram the Confederate ship, but she slipped away between the *Kennebec* and the *Ossippee.* At this moment the *Tennessee* put two shots into the *Ossippee* below her forward gun. Then, delivering a raking fire at the crippled *Oneida* as she went under that vessel's stern the *Tennessee* headed toward Mobile Point, apparently to take refuge under Fort Morgan's cannon.

As soon as the *Hartford* and *Metacomet* passed Fort Morgan the gunboat was cast loose to engage the unarmored Confederate vessels. She went first for the *Gaines,* damaged her badly with a shot under her counter, and drove her ashore where her crew set her on fire.

The *Morgan,* which had severely raked the *Hartford,* escaped from the *Metacomet* under cover of a rain squall and made her way safely to Mobile.

The *Selma* tried to get away from the *Metacomet* by crossing a shoal. As the Union ship followed the leadsman reported a foot less water than she drew. The captain, intent upon his quarry, said to the executive officer, "Call the man in; he is only intimidating me with his soundings."[6] A short time later the *Metacomet* overhauled and captured the smaller *Selma.*

After the Confederate gunboats had been dispersed Farragut sent his ships' crews to breakfast, intending to deal with the *Tennessee* at his leisure. The men had just begun to eat when (at 8:50 A.M.) a lookout discovered the ram bearing down upon the *Hartford*—contrary to what Farragut thought she was doing the *Tennessee* had not been trying to escape; instead she had been engaged in what was for her the cumbersome process of turning around. With breakfast forgotten, the Union ships gave their attention to the *Tennessee.*

The *Monongahela,* the first vessel to get under way, rammed the *Tennessee* on her starboard side. The crash almost wrecked the *Monongahela,* but did no damage at all to the *Tennessee.* Just before the collision occurred a couple of shots from the *Tennessee* wounded three men on the *Monongahela*'s berth deck; as the ships separated a 15-inch shot fired by the *Monongahela* hit the *Tennessee*'s casemate—it had about the same effect as a cork from a popgun would have had on a buffalo's skull.

6. A. T. Mahan, *The Gulf and Inland Waters* (New York: Charles Scribner's Sons, 1883), p. 329.

The *Lackawanna* next rammed the *Tennessee* hard enough to cause her to list considerably and to swing around until the two ships lay with their port sides almost touching. Most of the *Lackawanna*'s guns had been shifted to her starboard side the better to contend with Fort Morgan, but she had a 9-inch piece on the port side. A shot from this gun broke one of the *Tennessee*'s gun-ports, doing her the worst damage she had yet suffered.

The *Hartford* now tried to ram the *Tennessee,* but the latter sheered just far enough to prevent the pair from meeting head-on. However, they came close enough for one of the *Hartford*'s anchors to be bent as though it were a twig. As the ships passed each other a broadside of seven 9-inch guns left the *Tennessee* unscathed. When she attempted to reply most of her guns misfired; the one shot she did get off killed or wounded several men in the *Hartford.* As soon as the *Hartford* was clear she started to circle to get into position to ram the *Tennessee.* Before this maneuver could be completed the *Lackawanna,* also bent on ramming the Confederate vessel, collided with the *Hartford.* The blow cut the flagship down almost to the water line and Farragut, standing in the mizzen rigging, was fortunate to escape death. By the time the *Hartford* and *Lackawanna* were clear of each other and ready to try their luck again the slower moving monitors were closing the *Tennessee* and the fight was left for them to finish.

One of the *Manhattan*'s guns was useless because a piece of iron had dropped into its vent and could not be removed so she was able to fire only six shots, but one of them broke through the *Tennessee*'s casemate.

The *Winnebago*'s turrets were jammed, making it necessary to aim her guns by steering the ship; consequently her fire was extremely slow and she did no great damage to the *Tennessee.*

The *Chickasaw*'s smokestack had been hit several times and her speed had been reduced considerably. However, with tar and tallow heaped on her fires she overhauled the *Tennessee* and hung on doggedly astern of her, never more than 50 yards away, at times almost touching her, keeping up a relentless fire from her 11-inch guns. By this time the *Tennessee*'s bow and stern gunports had been jammed. Her smokestack was soon knocked down and smoke from its stump poured through the gratings onto the enclosed gun deck where the thermometer stood at 120 degrees Fahrenheit. Then her rudder chains were shot away and the captain was wounded by an iron splinter as he was supervising an attempt to repair the stern gunport. In these circumstances the executive officer surrendered the ship to prevent further loss of life.

The battle of Mobile Bay cost the Union Navy the loss of one monitor, 145 men killed, and 170 wounded; the Confederates lost three gunboats, had 12 men killed, and 21 wounded.

Once the fleet was in the bay Mobile was of no more use to blockade runners, but Farragut and Granger still wanted to capture Forts Morgan and Gaines.

On August 6 the *Chickasaw* and the gunboats bombarded Fort Gaines, which was surrendered the following day to avoid needless loss of life despite orders sent by signal to hold out as Fort Morgan intended to do.

Soon after Fort Gaines was captured Granger's troops landed three miles east of Fort Morgan and the Union ships, now including the *Tennessee,* began an attack on its northern face. Despite the strength of the forces arrayed against it the fort held out until August 23, by which time it was little more than a ruin.

When news of these events reached Washington on

Saturday, September 3 President Lincoln ordered a 100 gun salute to be fired the following Monday in honor of "the recent brilliant achievements of the fleet and land forces of the United States in the harbor of Mobile and . . . the reduction of . . . Fort Gaines and Fort Morgan."[7]

7. *NC,* pt. 4, p. 109.

14
Fort Fisher

Throughout the Civil War, Wilmington, North Carolina, situated on the Cape Fear River about 20 miles from the Atlantic Ocean, was one of the busiest seaports in the South; during the latter half of the war it was the Confederate States' principal port; after Farragut's fleet gained control of Mobile Bay, Wilmington was the only port available to blockade runners drawing 12 feet or more.

Wilmington was also the most difficult place to blockade on the Confederate States' coasts. This was so because there were two entrances to the Cape Fear River: one from the east called the New Inlet or Swash Channel, the other from the southwest called the Western Bar or Beach Channel. These passages are less than 10 miles apart as the crow flies, but with Smith's Island and Frying Pan Shoals lying between them they are separated by almost 40 miles of water. Because of this 30 or 40 vessels had to be kept constantly at those two stations, and they were unable to maintain as effective a blockade as 20 ships could at Charleston.

Welles recognized early in the war that Wilmington could not be strictly blockaded except by vessels stationed in the mouth of the Cape Fear River. Because he was aware that the Navy would need help to overcome the local defenses of the river before any ships could be kept there he sought in September 1862 "to get the consent of the War Department to a joint attack" on those defenses,[1] then consisting of only Fort Caswell, a masonry work on Oak Island commanding the Southwest Channel. The War Department felt that no troops could be spared for such an operation at the moment and President Lincoln was unwilling to interfere with the Army, even though he agreed with Welles about the importance of Wilmington, so the matter was shelved.

At or about this time the Confederates decided to strengthen the defenses of the Cape Fear River by building a fortification designed to guard the New Inlet Channel.

This work, called Fort Fisher, was situated on a peninsula between the Atlantic Ocean and the Cape Fear River named Federal Point in northern annals and Confederate Point in southern ones.

Fort Fisher was the most formidable military installation on the North American continent. It had two sod-covered earthen walls 25 feet thick at their bases, 8 to 12 feet thick at their tops, averaging 20 feet in height (the minimum was 8 feet; at one place, called the Mound Battery, a 10-inch columbiad and a 6⅜-inch rifle were mounted 60 feet above sea level). One of its walls, three-quarters of a mile long, faced the Atlantic Ocean; the other wall extended at a right angle from the first one for half a mile across the point almost to the Cape Fear River. A log palisade and a field of underground torpedoes (in modern terms, land mines) in front of the land face of the fort afforded a strong defense against an overland attack from the north. More than 40 of the

1. Daniel Ammen, *The Atlantic Coast* (New York: Charles Scribner's Sons, 1883), p. 215.

fort's 169 cannon were heavy pieces. And the Cape Fear River was too shallow to permit large vessels to run past the fort as they had been able to do at New Orleans and Mobile Bay.

Because Welles kept up his interest in a joint Cape Fear River expedition his subordinates did so too. However they realized, as did Welles, that the problem had been made far more difficult by the construction of Fort Fisher.

In May 1863 Admiral Lee reported that his ships could not occupy the river until Forts Caswell and Fisher had been overcome, a task which he admitted the Navy could not accomplish without the assistance of an adequate land force.

Late the following September, Commander William A. Parker, who had made an independent survey of the two forts, wrote:

> I am of the opinion that 25,000 [soldiers] and two or three ironclads should be sent to capture this place, if so large a force can conveniently be furnished for the purpose. . . . The ironclads should be employed to divert the attention of the garrison at Fort Fisher during the landing of our troops at Masonboro Inlet, and to prevent the force there from being used to oppose the debarkation.[2]

The War Department was as unwilling at this time as it had been a year earlier to furnish enough troops for the sort of expedition Welles had in mind. However, he brought the matter up again in 1864 after Farragut's victory at Mobile Bay. He did so the more vigorously because Wilmington was now the only major port open to blockade runners and it seemed to him, as he said later, that to close it, "and thus [to] terminate the inter-

2. *Civil War Naval Chronology* (Washington, D. C.: Naval History Division, Office of the Chief of Naval Operations, Navy Department, n. d.), pt. 3, p. 158. (Hereafter cited as *NC.*)

course of the [Confederate States] with the outside world, would be like severing the jugular vein in the human system."[3]

That September he undertook to convince the Secretary of War that the moment was fit for an expedition against Fort Fisher because with "the armies . . . mostly going into winter quarters" there should be plenty of troops available.[4]

At the same time he suggested to Grant, who was now general in chief of the Union Armies, that if Wilmington were cut off the Confederates would have to abandon Richmond, then besieged by the Army of the Potomac. This happy thought induced Grant to have General Weitzel reconnoiter Fort Fisher. Weitzel's report convinced Grant of the futility of the proposed movement, but Welles kept insisting on it until Grant at last gave in.

As soon as Grant agreed to provide the troops necessary for a Fort Fisher expedition an order was sent to Farragut to take command of the North Atlantic Squadron which would furnish the naval contingent for it. Before this order reached Farragut he had written to the Navy Department asking to be relieved because his health was deteriorating. In these circumstances the duty was assigned to Admiral Porter who assumed command of the squadron on October 1.

A force of 150 vessels of all sorts, including the *New Ironsides,* a double turretted monitor, and three standard monitors was assembled at Port Royal and put at his disposal.

While plans for the expedition were being made General Butler, now commanding the Army of the James, read a newspaper account of the great destruction caused

---

3. Gideon Welles, "Lincoln's Triumph in 1864," *The Atlantic Monthly* 41 (April 1878): 459.
4. *NC,* pt. 5, p. 5.

in England by the accidental explosion of 75 tons of gun powder in a canalboat. This led him to suggest that if a shipload of gunpowder were exploded close aboard Fort Fisher's ocean face the garrison might be demoralized enough to permit troops landing immediately afterward to take the works without a great deal of difficulty.

On November 23, 1864, a group of Army and Navy ordnance experts who met in Washington to consider Butler's idea decided that the explosion of 300 tons of gunpowder in a vessel beached near Fort Fisher

> would injure the earthworks to a very great extent, render the guns unserviceable for a time, and probably affect the garrison to such a degree as to deprive them of the power to resist the passage of naval vessels by the fort and the carrying of these works by immediate assault.[5]

In view of this decision the Navy Department had a shallow-draught gunboat, the *Louisiana,* a former merchant steamer, converted into a waterborne bomb. Although a few persons expressed some doubts, the consensus seems to have been that the powder boat, as it came to be called, would be a dreadfully destructive thing —almost comparable to a modern atomic bomb. Among those who were most optimistic about it were Fox and Porter. The latter, who had proposed a somewhat similar method of destroying a fort during the Mexican War, thought the explosion of the *Louisiana* "would wind up Fort Fisher and the works along the beach," thus eliminating all risk to the fleet. He officially informed Butler that human ingenuity could not have devised a more formidable weapon than had been made out of the *Louisiana.* His sincerity is demonstrated by his actions. When it came time for the powder boat to be exploded he

5. *Official Records of the Union and Confederate Navies in the War of the Rebellion* (Washington, D. C.: Government Printing Office, 1894–1922), Ser. 1, 11: 216. (Hereafter cited as *ORN.*)

ordered his ships to run 12 miles out to sea and blow off their boilers to prevent them from being badly damaged and he told the officer who had charge of the boat that he expected "more good to come to [the Union] cause from a success in this instance than from an advance of all the armies in the field."[6]

The experts who supervised the preparation of the *Louisiana* agreed that in order to produce the maximum effect from the blast it would be necessary to explode the entire cargo of gunpowder as nearly instantaneously as possible, but they were puzzled about how to accomplish this result. Electricity, they concluded, was too unreliable an agent. Accordingly they decided to use several clockwork devices to drop weights upon some percussion caps set in a tub full of gunpowder with Gomez fuses (rubber tubing filled with fulminating powder) leading from the tub to the gunpowder stowed in the ship. As a precaution against a possible failure of these mechanisms a slow match, to be lighted when the clocks were started, was led into the tub. To insure an ultimate explosion so that the cargo would not fall into the Confederates' hands the ship was to be set on fire by the men who beached her just before she was abandoned.

Because Fort Fisher and Wilmington were situated within Butler's military department General Grant instructed him to select the troops for the expedition and to put them under the immediate command of General Weitzel. Butler, who was eager to witness the powder boat experiment, persuaded Grant to let him accompany the troops. Thus it came about that the joint commanders of the Fort Fisher expedition were men who were barely on speaking terms because Porter had never forgiven Butler for his factual statement about the inconsequential results of the mortar bombardment of Forts St. Philip

6. Ibid., Ser. 1, 11: 217, 219; *NC,* pt. 4, p. 147.

and Jackson. And Butler's second in command was someone Porter hated quite as venomously as he did Butler because Weitzel, too, had spoken slightingly, albeit accurately, of the effect of the mortar fire.

If the term "snafu" had been current at the time of the Civil War it could most appropriately have been applied to the Fort Fisher expedition. Although the naval contingent of the joint force was ready to depart in mid-October, Grant waited almost until the end of November before he assigned 6500 troops (less than a quarter of the number Commander Parker thought necessary) to the expedition. Then, when the transports finally left Hampton Roads, they sailed directly to the rendezvous, off New Inlet, while the naval vessels went first to Beaufort to take on ammunition and to finish loading the *Louisiana*'s cargo of 315 tons of gunpowder. Because the Army had already delayed the movement for a long time according to Porter's way of thinking he did not feel it necessary for him to hurry away from Beaufort. As a result of his slowness the transports waited three days for the warships to join them. The weather during this time was ideal for the swimming and fishing the soldiers did to improve their enforced leisure; it would have been equally suitable for an attack on Fort Fisher.

When Porter arrived with his fleet late in the afternoon of Sunday, December 18, he decided, *without consulting Butler,* to commence operations at once. About 8 P.M. he notified Butler in writing that the powder boat was on its way ashore to be exploded at 9:20 P.M. Butler, Weitzel, and Colonel Cyrus B. Comstock (of Grant's staff, who was present as an observer) argued vigorously that this schedule would give the Confederates the whole night to repair any damage that might be done to the fort. With some difficulty they persuaded Porter to recall the *Louisiana* and by luck a fast tug overhauled her before it was too late.

Monday dawned so stormy as to make it obvious that no landing could even be attempted for several days at least. Therefore, Butler, with Porter's concurrence, left for Beaufort with the transports to wait for the gale to subside and to replenish the ships' supplies of coal and water which were running low.

Butler sent word to Porter from Beaufort that the transports would rejoin the warships at sunset, December 24. *After receiving this message* Porter decided to explode the powder boat at 1:20 A.M., December 24. Although he knew exactly when he could expect the transports to return, his official explanation to his superiors for his conduct in this matter was that he felt he could not afford to lose the spell of good weather following the three days' storm "and the transports with the troops not making their appearance [he] determined to attack Fort Fisher and its outworks."[7]

To the best of the writer's knowledge and belief no historian in any way connected with the Navy has ever questioned Porter's statement that he acted as he did because he was uncertain when the troops would return, even though some of them must have known the truth.

However, a contemporary soldier voiced a not illogical suspicion that Porter caused the powder boat to be exploded when he did because he expected it to do so much damage to the fort that a few sailors could step ashore and accept the surrender of its survivors, thus gaining great kudos for himself and the Navy at the expense of Butler and the Army.[8]

Although the importance of following the explosion of the powder boat with an immediate attack had been repeatedly stressed, the fleet did not stand in toward shore until daybreak and did not open fire on the fort until 11:20 A.M., about 10 hours after the boat was blown up.

---

7. Ibid., Ser. 1, 11: 247.
8. H. C. Lockwood, "The Capture of Fort Fisher," *The Atlantic Monthly* 27 (May 1871): 627.

Thus if the fort had been damaged ample time would have been available for its repair and if the garrison had been affected there would have been plenty of time for its recovery.

When the naval bombardment finally began the fort scarcely bothered to reply to it in order to conserve ammunition. Nevertheless, the Confederate fire was brisk enough to kill eight sailors and wound 57 others. (Sixteen men were also killed and 27 wounded by the bursting of Parrott rifles in five ships.)

Porter optimistically concluded that the fort's desultory fire was the result of the fleet's gunnery and that as soon as the troops arrived they could occupy the place without meeting any opposition. He reported to the Navy Department that because of the "severe bombardment that followed" the explosion of the powder boat "200 men could have gone in and taken possession of the works."[9]

Even if this estimate of the situation had been correct Porter knew that the nearest Union soldiers were at Beaufort, 70 miles away, and that none would reach the vicinity of the fort for several hours.

Actually he could hardly have made a less accurate estimation of the condition of the fort. The explosion of the powder boat did not damage it at all and, far from demoralizing the garrison, the blast was not even heard on shore except by men who were awake when it occurred. An officer who did hear it thought a Union gunboat had blown up some distance away. The fort's commanding officer believed a blockade runner had been driven ashore and been destroyed by her own crew. He made no mention of the incident in his detailed report to the Confederate War Department about the attack on the fort. And the casualties resulting from the ships' "severe bombardment" were one man mortally wounded, three severely wounded, and 19 slightly wounded.

---

9. *ORN*, Ser. 1, 11: 261.

When the powder boat experiment turned out to be a complete failure everybody connected with the Union Navy claimed (none louder than Fox and Porter) never to have had any faith in it from the beginning. However, their words and actions had gone into the records where they could not be erased, even though they preferred to forget what they had said and done.

Before the transports were ready to leave Beaufort, Butler learned that the powder boat had been exploded. Immediately after he returned to the vicinity of Fort Fisher, at the time he had promised he would, he sent a staff officer to the flagship to arrange for a conference with Porter. The Admiral sent back word "that he was too much fatigued to give them an audience, but [he] would receive General Weitzel and Colonel Comstock in the morning."[10]

When Porter "received" Weitzel and Comstock they found him supremely confident that Fort Fisher had been practically destroyed. However, they managed to persuade him to bombard it again and to provide a covering fire for the landing of the troops.

At first the troops met with no resistance from the Confederates or trouble from the elements. However, the sea began to grow rough toward noon and by 3 P.M., when 2500 to 3000 men had gotten ashore operations had to be suspended because so many boats had been upset.

By this time Weitzel, in command of the landing party, had captured some Confederates in an outlying earthwork who, to his surprise, were from a division he knew had recently been in the trenches near Richmond. Soon afterward he took some more men who said they had been sent outside of the fort because there was no room for them in it. These events convinced him that the place

10. Lockwood, "Capture of Fort Fisher," p. 630.

had been heavily reinforced. Examining the fort through a telescope at a distance of about 150 yards he saw that the naval bombardment had done practically no damage to it. Besides counting 16 big guns facing the direction from which the troops would have to advance he saw that the fort's grass covered walls were wholly unbroken. With Comstock's concurrence, Weitzel reported to Butler that it would be suicidal to attack the fort with the troops available at the moment.

About the time Weitzel's report reached Butler he learned that the warships had used up most of their ammunition and he was advised by one of his subordinates, a former naval officer, that the men on shore would either have to be furnished with provisions enough for several days or be taken off because the barometer was falling rapidly, indicating an approaching storm. After satisfying himself of the accuracy of Weitzel's description of the condition of the fort and considering all of the factors involved *except that Grant had ordered the troops to hold any beachhead they gained* Butler wrote to Porter: "I see nothing further that can be done by the land force. I shall, therefore, sail for Hampton Roads as soon as the transport fleet can be got in order."[11]

Commenting on Butler's decision to withdraw the troops, Porter claimed that Fort Fisher had been "so blown up, burst up, and torn up that the people in it had no intention of fighting any longer" and that "there never was a fort that invited soldiers to walk in and take possession more plainly than Fort Fisher."[12]

As a naval officer who knew Porter well once said, if he believed anything he "never thought it worth while to inquire as to the facts."[13] In the case of Fort Fisher the

---

11. *ORN,* Ser. 1, 11: 251.
12. Ibid., Ser. 1, 11: 261–62.
13. Daniel Ammen, *The Old Navy and the New* (Philadelphia: J. B. Lippincott Company, 1891), pp. 461–62.

truth was (as Weitzel said it was) that an assault would have been suicidal. The commanding officer of the fort wrote later:

> When Butler's skirmish-line approached I purposely withheld the fire of the infantry and artillery until an attack should be made in force. Only one gun on the land-face had been seriously disabled, and I could have opened a fire of grape and canister on the narrow beach which no troops could have survived.[14]

Butler's failure to take Fort Fisher did not greatly disturb Grant, but he was outraged about the withdrawal of the troops despite his clear, nondiscretionary order for them to hold any foothold they secured.

After reading Butler's report about the expedition Grant asked the Secretary of the Navy to keep Porter's fleet in the vicinity of Fort Fisher until fresh troops and another general could be sent there. At the same time he reported to the Secretary of War that the debacle had been largely caused by the Navy's tardiness, because during the three days of good weather which elapsed after transports reached the scene of action and the time the naval vessels arrived there the troops might have taken the fort while it had too few men to hold it, but the delay had given the Confederates time to accumulate a strong force.[15]

Although the Navy Department never publicly made any direct comment on Grant's implied criticism of Porter's conduct, Fox did say privately to Porter while preparations were being made for the second Fort Fisher expedition that the country would not forgive another failure.[16]

Porter, obviously trying to exculpate himself, wrote to

---

14. Robert U. Johnson and Clarence Buel (eds.), *Battles and Leaders of the Civil War* (New York: The Century Co., 1884–87), 4: 646, ftn.
15. *ORN,* Ser. 1, 11: 357–58.
16. Ibid., Ser. 1, 11: 409.

Grant: "Please impress the commander [of the troops] with the importance of consulting with me fully as regards weather and landing"—[17] this in the face of his own studied lack of cooperation with Butler.

On January 12, 1865, 8000 troops, under the command of General Alfred S. Terry, sailed from Hampton Roads bound for Fort Fisher. At the same time a siege train of twenty 30-pound Parrott rifles, four 100-pound Parrotts, and twenty coehorns was sent to Beaufort to be held subject to Terry's orders.

At 4 A.M. the following day the naval vessels opened fire on the fort. Four hours later the troops began landing in nearly 200 small boats assisted by tugs. By midafternoon all of the soldiers had been landed, each man carrying 40 rounds of ammunition and rations for three days. Before nightfall an additional 300,000 rounds of small arms ammunition, six days' supply of hardtack, and a large number of entrenching tools had been put ashore.

Terry quickly concluded that landing siege equipment on the open beach would be extremely difficult if it could be done at all. Therefore, he decided to attempt to carry the fort by assault provided that by January 15 the naval vessels could destroy enough of the palisade north of the land face to make such a move practicable. Porter was informed of this decision and he promised to do his best.

On this occasion the Navy was far more helpful to the Army than it had been in the earlier attack. The four monitors stood in at least a quarter of a mile closer to the fort and in a much better position to enfilade the land face which the troops would have to assault. The unarmored vessels also moved in closer and concentrated their fire upon the land face. The ships' guns were aimed with far more care than they had been during the previous

17. Ibid.

attack. And the bombardment was kept up all night, making it impossible for the Confederates to repair the fort and difficult for them to bury the dead.

(Considering how much more helpful the Navy was to Terry than it had been to Butler one is almost forced to conclude that Porter wanted the first expedition to fail, if he did not actually sabotage it.)

Perhaps because he was resentful of the justifiable criticism of the Navy's poor showing in connection with the December expedition Porter detailed 1600 sailors and 400 marines to go ashore armed with cutlasses and revolvers "and board the fort in a seaman-like way." His order said, "The Marines will form in the rear and cover the sailors. While the soldiers are going over the parapets [of the land wall], the sailors will take the sea face of Fort Fisher."[18]

This order was more easily issued than executed, especially since the landing party was composed of men from 35 vessels who had been on shipboard for months and had never drilled together. To make things worse nobody in the party knew when it landed who was in command of it nor to whom the officers from the various ships were supposed to report.[19]

After the senior officers had more or less straightened out the command problem the men marched along the beach to a point just out of musket shot of the fort where they lay down to wait for the signal to charge. When it was given, by the blowing of all the ships' whistles at 3 P.M., they started a rush toward the fort and ended up "packed like sheep in a pen, while the [Confederates] were crowding the rampart not forty yards away, . . . shooting . . . as fast as they could fire,[20] killing at least 80 and wounding nearly 300 of them.

---

18. *NC,* pt. 5, p. 6.
19. Ammen, *The Atlantic Coast,* pp. 232, 238–39.
20. *BL,* 4: 660.

However, their long lines and numerous flags had presented such a formidable appearance that the fort's commanding officer was led to think they were going to make the main attack. While they held the attention of most of the garrison the soldiers gained a foothold at the other end of the land face. The men of both sides fought doggedly until the last of the fort's defenders were overcome shortly before midnight.

After losing Fort Fisher the Confederates abandoned Fort Caswell and some minor works at and near the mouth of the Cape Fear River.

Early in February, Grant visited Porter in his flagship anchored off Fisher to plan with the Admiral for a joint movement up the Cape Fear River to Wilmington.

As soon as Grant returned to his headquarters he ordered an army corps to be transferred by sea from Alexandria and Annapolis to Fort Fisher. In the meantime Porter, in anticipation of his next move, sought to have the three monitors he had sent back to Charleston returned to him. However, because Welles and Fox were so obsessed with Charleston, Dahlgren was able to persuade them that he needed the vessels more than Porter did.

While Porter was waiting for the troops to arrive he engaged Fort Anderson, on the west bank of the Cape Fear River halfway between Fort Fisher and Wilmington, to test its strength. On the basis of what he learned he issued an operations order reading:

> The object will be to get the gunboats in the rear of their [the Confederates'] intrenchments and cover the advance of our troops. When our troops are coming up, the gunboats [are to] run close in and shell the enemy in front of them, so as to enable the troops to turn their flanks, if possible. . . . As the army come[s] up, your fire will have to be very rapid, taking care not to fire into our own men. . . . Put yourself in full

communication with the commanding general on shore, and conform in all things to his wishes.[21]

The last of the troops reached Fort Fisher on February 8. When they were ready to move on February 16 the gunboats helped to ferry them to Smithville on the west bank of the river.

Two days later Fort Anderson was overcome by 8000 troops assisted by four gunboats and the monitor *Montauk*. Thereupon the Confederates fell back to Town Creek and Sugar Loaf Hill on the east bank of the river about three miles below Wilmington. Here, on the night of April 21-22, a desperate attempt was made to stop the Union advance by sending a couple of hundred floating torpedoes down the river. The crews of the monitor and the gunboats managed to fend off most of them, but the *Osceola* had one of her side-wheels and a paddle box damaged.

By noon April 22 Wilmington was occupied, Scott's Anaconda plan was at last fulfilled, and, as Welles had foreseen, the capture of Wilmington brought an end to the Civil War.

21. *NC,* pt. 5, p. 37.

15

Confederate Cruisers

Although the commerce raiding done by its seven cruisers was the most fruitful of the Confederate Navy's operations, it still had no effect on the outcome of the Civil War because Welles wisely refused to detach ships from the blockading squadrons to hunt for them. However, their histories are interesting enough to deserve a brief description.

The first of these craft, the *Sumter,* was originally the *Habana,* a 500-ton bark-rigged steamer plying between New Orleans and Havana. She was converted into a warship of sorts by having her frames strengthened, her spar deck passenger cabins removed, a berth deck and magazines installed, and her bunkers enlarged enough to carry an eight-day supply of coal. Her armament consisted of an 8-inch pivot gun mounted between the fore- and mizzenmasts and four 24-pounders in broadside.

On April 18, 1861, Captain Raphael Semmes took command of the *Sumter* at New Orleans. He hoped to get her to sea quickly, but was unable to do so before the U.S.S. *Brooklyn, Massachusetts,* and *South Carolina* began blockading the mouths of the Mississippi River.

When the Sumter was finally ready to sail, on June 18, she dropped down to the Head of the Passes to wait for an opportunity to slip out. Semmes, of course, had every advantage in this game. He could keep informed of the movements of the Union ships while their captains were ignorant even of the *Sumter*'s existence.

On June 30 the *Brooklyn* left her station at the mouth of Pass a l'Outre to investigate a strange sail and the *Sumter* hurried into the Gulf. She was sighted too late to be headed off before she crossed the bar. The *Brooklyn* chased her under canvas and power, but the *Sumter* escaped by heading in such a direction that the *Brooklyn* could not use her sails. (Under steam alone the *Sumter* was the faster vessel.)

Three days later the *Sumter* captured and burned her first prize, the bark *Golden Rocket,* off the coast of Cuba near the Isle of Pines. On the last day of her first week at sea the *Sumter* escorted seven prizes into Cienfuegos, Cuba. Semmes put these craft into the hands of a local agent and told the Governor that he had come into port "with the expectation that Spain will extend to the cruisers of the Confederate States the same friendly reception that in similar circumstances she would extend to the cruisers of the enemy. . . ."[1] To Semmes's bitter disappointment his prizes were soon released by the Spanish authorities.

During the next two months the *Sumter* cruised in the Caribbean Sea and along the coast of South America. She was received in as friendly a manner as Semmes could possibly have wished and allowed to stay as long as he cared to do at most ports. She coaled without hinderance at Curaçao, the West Indies; Paramaibo, Guiana; and Maranhao, Brazil; but she was permitted to stay only 48 hours in Puerto Cabello, Venezuela.

After spending a couple of months in the Atlantic the *Sumter* put into St. Pierre, Martinique, for coal and water.

---

1. *Civil War Naval Chronology* (Washington, D. C.: Naval History Division, Office of the Chief of Naval Operations, Navy Department, n. d.), pt. 1, p. 18. (Hereafter cited as *NC.*)

She was discovered there by Captain James S. Palmer, commanding the U.S.S. *Iroquois,* when he happened to come into St. Pierre soon after the *Sumter* did. On being warned by the French authorities that the rule forbidding one belligerent vessel from sailing with 24 hours of another would be strictly enforced Palmer departed immediately but he stood off and on the 12-mile-wide roadstead as close in with the shore as he could without risking a charge of hovering in neutral waters.

Palmer arranged for an American schooner lying in the harbor to signal when the *Sumter* sailed and in what direction she was heading. Semmes, of course, suspected that something of this sort would have been done and planned to turn it to his own advantage. When the *Sumter* was ready to depart, during the night of November 23, she headed south until Semmes was sure the *Iroquois* was following her; then she reversed her course. By the time Palmer became aware of Semmes's ruse the *Sumter* was too far away to be overhauled.

Cruising to the eastward the *Sumter* took and burned three more prizes before bad weather forced her to put into Cadiz, Spain, early in January 1862. After a brief stay there she went to Gibraltar where she was soon found by the U.S.S. *Tuscorora.* When that ship was joined by the U.S.S. *Ino* and *Kearsarge,* Semmes gave up all hope of getting the *Sumter* to sea again and sold her.

Long before the *Sumter*'s cruise ended the Confederate government decided that rather than trying to build vessels suitable for commerce raiding in the South it would seek to procure them abroad. Accordingly several agents, headed by Captain James S. Bulloch of the Confederate Navy, were sent to Great Britain for this purpose.

As soon as Bulloch reached England, in June 1861, he sought advice on how to evade the British Foreign

Enlistment Act, which, among other things, forbade the Queen's subjects building, arming, or equipping vessels of war for use against nations friendly to Great Britain. He was told that nothing in the Act made it illegal to construct a ship as one operation and nothing in it prohibited the sale of arms and munitions suitable for such a vessel as a separate transaction. And, provided it happened more than three nautical miles from the British coast, a ship could be equipped with arms and supplied with ammunition without violating the Act. This Jesuitical construction of the statute was propounded in Parliament by legal representatives of the Crown and was formally laid down for the guidance of juries by the Lord Chief Baron of the Court of the Exchequer.[2] (Obviously the lawyers and legislators who thus interpreted the Foreign Enlistment Act did so with a conscious view to permitting the shipping of Great Britain's chief commercial rival to be hurt severely, if not destroyed.)

Assured of a free hand, Bulloch lost no time in making a contract with William C. Miller and Sons for the construction of a steam barkentine, named the *Oreto,* to be built to the specifications of a British cruiser.

Twice during the winter of 1861-62 the United States minister to Great Britain, Charles Francis Adams, called the attention of the Foreign Office to his strong suspicions that the *Oreto* was being built for use as a Confederate warship. On both occasions *pro forma* inquiries were made, but the authorities declared themselves unable to find that any British law was being violated.

On March 22, 1862 the *Oreto* sailed from Liverpool under the ostensible ownership of a local merchant and the actual command of a British captain. At the same time the steam bark *Bahama* left Hartlepool, bound for

2. Charles Francis Adams, *Charles Francis Adams* (Boston: Houghton, Mifflin and Company, 1900), pp. 307–8.

Nassau, with two 7-inch rifles and six smoothbores in her hold.

The *Oreto* was delivered to Adderly and Company, a Confederate agent in Nassau, on April 28. When the *Bahama,* also consigned to Adderly and Company, arrived at Nassau a few days later her cargo and the *Oreto* were transferred to Commander John N. Maffitt of the Confederate States Navy.

With more audacity than discretion Maffitt began transshipping the *Bahama*'s cargo, including the cannon, to the *Oreto* in the harbor at Nassau. At the insistence of the United States consul the captain of a British naval vessel was ordered to inspect the *Oreto*. His report that she was in every way fitted as a man-of-war led to her being libelled in the vice-admiralty court. After a trial in which the court's sympathy with the libellee was scarcely concealed, the *Oreto* was released on August 7. She then sailed a short distance to Green Cay where Maffitt commissioned her as C.S.S. *Florida*.

Because rammers, beds, locks, and quoins for the *Florida*'s guns had carelessly been omitted from the *Bahama*'s cargo and only 22 of the men who brought the *Oreto* to Nassau were willing to serve in a Confederate warship Maffitt sailed for Cuba where he hoped to recruit a crew and obtain equipment. After spending a fruitless week in Cuba he decided to attempt to reach Mobile despite the fact that he and most of the *Florida*'s crew had contracted yellow fever.

Knowing the *Florida* to be a duplicate of several British warships which had been cruising on the southern coasts and inspecting the efficiency of the blockade Maffitt approached Mobile Bay in midafternoon, September 4, with his ship flying a British flag. Because foreign naval vessels always communicated with the senior officers of local blockading squadrons before entering southern ports the two Union ships stationed off Mobile held their fire until it finally became certain that the *Florida* was not going to do the usual thing. Thus, making 14 knots to their seven, the *Florida* ran into Mobile Bay without suffering much damage from the Union ships' gunnery.

In the hope of keeping the *Florida* bottled up in Mobile five more Union vessels were sent there, but she slipped past them on January 16, 1863.

Ten days after leaving Mobile the *Florida* put into Nassau to land the crews of three prizes she had taken and burned because there was no hope they could reach any Confederate port. Her arrival there as an undisguised man-of-war apparently did not embarrass the authorities who had allowed her to be delivered to Maffitt as a merchantman not too long ago. She was warmly greeted, permitted to stay in port for 36 hours instead of the 24 hours prescribed by international law for a visiting warship of a belligerent power, and furnished with coal enough to last her for three months, again in flagrant disregard of the rule that such a vessel should be supplied with only enough fuel to reach the nearest port in her own country.

During the next five months the *Florida* took 14 prizes while cruising from the latitude of New York to that of Bahia, Brazil.

At the request of Lieutenant Charles W. Read, Maffitt armed one of the captured vessels, the brig *Clarence*, with one gun, furnished her with a crew of 20 men, and sent her on an independent cruise under the Lieutenant's command.

Read originally intended to use the *Clarence*'s papers to bluff his way into Hampton Roads and burn as many as he could of the Union merchantmen he thought would be congregated there. When he learned from the people of a couple of prizes he took off the coast of Virginia that there were only a few merchant vessels at Hampton Roads and the place was well guarded he decided to cruise to the northward, doing as much damage as he could.

In June the *Clarence* took five prizes between Chesapeake Bay and the vicinity of Portland, Maine. The last of these vessels, the schooner *Tacony,* was such a fine sailer that Read kept her and burned the brig.

During the next 10 days the *Tacony,* renamed *Florida No. 2,* captured 15 vessels of various sorts. Then, to hide his tracks, Read burned the *Florida No. 2* and armed the last of his prizes, the fishing schooner *Archer.*

Early in the morning of June 27 the *Archer* was piloted into Portland by two unsuspecting fishermen from whom Read learned that the Revenue Cutter *Caleb Cushing* was in port. Read's original plan had been to set fire to the shipping in the harbor; now he decided to seize the cutter. After the pilots were dropped the *Archer*'s people boarded the *Caleb Cushing,* overpowered her crew, and put them below in irons.

On the morning of June 28 the residents of Portland saw the familiar *Caleb Cushing* and a strange vessel barely visible in the distance. Two steamers and a tugboat were hastily manned by soldiers from nearby Fort Preble and sent off at full speed to find out what game was afoot. They overhauled the *Archer* about 20 miles at sea, identified her as a Confederate vessel, and took her after her ammunition was exhausted. The *Caleb Cushing* was recaptured later the same day.

When the *Clarence* and the *Florida* separated the *Florida* headed for Bermuda where she was refitted and supplied with coal. She sailed from there for Brest, France, where she stayed for nearly six months.

Early in February 1864, the *Florida,* now commanded by Lieutenant Charles M. Morris, eluded several Union ships watching for her outside of Brest and began an Atlantic cruise which lasted until she put into Bahia the following October 4.

As it happens Commander Napoleon Collins, commanding the U.S.S. *Wachusett,* who had been hunting for the *Florida* for several months, saw her enter Bahia. He followed her into the harbor and challenged Morris to meet him offshore in a ship duel. The Confederate captain wisely refused this proposal and, relying on the 24-hour rule, hoped to give the *Wachusett* the slip. When his challenge was refused Collins decided to prevent the *Florida* from escaping by destroying her where she was.

The Brazilian authorities, who recognized the explosive nature of the situation, extracted promises from Morris and the United States consul that neither vessel would attack the other while they remained in port.

Despite the consul's promise the *Wachusett* slipped her cable, steamed around a Brazilian corvette lying between her and the *Florida,* and rammed the *Florida* on her starboard quarter at about 3 A.M., October 7. The collision cut the *Florida* down to her bulwarks and knocked down her mainmast, but did not sink her. The *Wachusett* then backed clear and opened fire with her heavy guns. Since Morris and most of the *Florida*'s crew were on shore her acting commander had no choice except to surrender her. She was thereupon towed out of the harbor and convoyed to Hampton Roads with a prize crew on board.

"The capture of the *Florida,*" as James R. Soley wrote, "was as gross and deliberate a violation of the rights of neutrals as was ever committed in any age or country."[3] Brazil, of course, protested Collins's brazen act. He was court-martialled and the United States

---

3. James Russell Soley, *The Blockade and the Cruisers* (New York: Charles Scribner's Sons, 1883), p. 189.

promised to return the *Florida* to Brazil so that she could be restored to the Confederate States. The court-martial found Collins guilty and sentenced him to be dismissed from the Navy. However, Welles, who knew little and cared less about international law, set the verdict aside and restored Collins to duty. Soon after it was promised that the *Florida* would be returned to Brazil she was rammed and sunk by an Army transport at Hampton Roads. Two courts of inquiry investigated this event, but they were unable, or unwilling, to decide whether it was a real accident or a contrived one.

The second cruiser built for Bulloch was the famous *Alabama*.

This vessel, known on the stocks as No. 290 because hers was the 290th keel laid at the John Laird and Sons shipyard in Birkenhead, near Liverpool, was launched in the spring of 1862 as the *Enrica*.

The *Enrica,* a handsome steam barkentine built to the specifications of a British cruiser, was designed more for speed than for mounting a heavy battery. She had long lower masts, enabling her to carry a big spread of canvas, and two 300 horsepower engines arranged to drive one propeller which could be lifted out of the water to eliminate its drag when it was not in use. With a speed of 10 or 12 knots under sail alone and 15 knots using sail and steam together the *Enrica* could overhaul or run away from almost any other vessel afloat.

About the time the *Enrica* was launched Adams presented indisputable evidence to the British Foreign Secretary that she was intended for service against the United States. The law officers of the Crown simply could not refuse to act in the light of Adams's information, but they managed to move so slowly there was time for somebody to tell Bulloch "it would not be safe to leave the ship at Liverpool another forty-eight hours."[4]

4. Adams, *Charles Francis Adams,* pp. 314–15.

Thus warned, Bulloch quickly arranged to get the *Enrica* to sea. To conceal his real purpose he invited a large number of local socialites to enjoy a day's sail with him in the new vessel. Soon after she left the harbor the guests were sent ashore in a tugboat and the *Enrica* made her way to Point Lymas about 50 miles from Liverpool. She stayed there for two days, untroubled by the British authorities, then steamed up the Irish Channel, rounded the northern coast of Ireland, and disappeared into the Atlantic Ocean.

On Sunday, August 10, the *Enrica* arrived at Porto Praya on the island of Terceira, the Azores. Eight days later the *Agrippina,* a bark Bulloch had bought, came into Porto Praya carrying a 100-pound Blakely rifle, an 8-inch smoothbore, and six 32-pounders, ammunition, coal, and provisions enough for a long cruise. The *Enrica* and the *Agrippina* immediately departed for the little frequently Bay of Angra, Terceira. That afternoon they were joined by the *Bahama* with Captain Semmes and 25 other Confederate naval officers on board. By 10 P.M. the following Saturday the *Agrippina*'s cargo had been transferred to the *Enrica,* which had the Blakely rifle pivoted on her forecastle, the 8-inch piece pivoted abaft the mainmast, and the 32-pounders mounted in broadside. At noon Sunday the *Enrica* was commissioned by Semmes as the C.S.S. *Alabama* while the band played *Dixie.*

Attracted by the hope of prize money, which Semmes promised to pay even for vessels destroyed, 81 British seamen and one officer, Dr. D. H. Llewellyn, enlisted as the *Alabama*'s crew.

The *Alabama* took her first prize, the *Ocmulgee,* an Edgartown, Massachusetts, whaleship, 11 days after being commissioned. Her 21st prize, taken December 7, 1862, was the steamer *Ariel,* bound from New York for Aspinwall, New Grenada (in what is now the Republic of Panama), with mail, freight, and about 700 passengers on board. Semmes originally planned to land the passengers and crew at Kingston, Jamaica, then to burn the ship at sea. On learning that there was an epidemic of yellow fever at Kingston he released the *Ariel* under a ransom bond to be paid when the war ended, victoriously for the Confederate States as he expected it would.

Semmes deduced from some newspapers he found in the *Ariel* that a Union expedition commanded by General Banks was en route to Galveston. (Actually Banks was on his way to relieve General Butler at New Orleans.)

On the basis of the evidence available to him Semmes reasoned that Banks's force would reach Galveston on January 10, 1863. Accordingly the *Alabama* was coaled at a rendezvous with her tender at an island off the coast of South America and she departed for Galveston on January 5.

Planning to spot the Union fleet before his ship was seen, Semmes arrived at Galveston about 3 P.M., January 11. He expected to come upon a large number of transports outside of the bar, hoped to sink or set fire to many of them, and to escape before a superior force could pursue him.

Instead of finding a collection of nearly helpless transports the *Alabama* suddenly came upon five Union warships blockading Galveston. Ordinarily Semmes would have beaten a hasty retreat from so powerful a force. However, the carelessness of the men at the *Alabama*'s masthead led to her being seen before Semmes, on deck, realized he was running into a hornets' nest.

Fortunately for the *Alabama* the lookout in the U.S.S. *Brooklyn,* who first saw her, mistook her for a three-masted schooner. Ordinarily the ship which discovered a stranger intercepted it in the hope of gaining some prize money, but the *Brooklyn* was having new grate bars installed and the task was assigned to the *Hatteras,* a former Delaware River side-wheeler—it did not seem necessary to send more than one vessel to check on a schooner.

When the *Hatteras*'s lookout recognized the chase as a steamer, the fact was reported by signal, but none of the other Union ships read the message.

For a while the *Hatteras* gained slowly on the stranger. When the distance between them suddenly began to lessen, at about 6:30 P.M., the commanding officer of the *Hatteras,* Lieutenant Commander Homer C. Blake, became suspicious enough to have his ship cleared for action. By this time it was quite dark, but the other vessel's appearance and actions led him to believe she was the *Alabama.*

Because the *Hatteras* could work only two short 32-pounders, a 30-pound Parrott rifle, and a 20-pound rifle on a side Blake waited until he was close aboard the stranger before he hailed her. Finally he called: "What steamer is that?" The answer he thought he heard was: "Her Britannic Majesty's Ship *Vixen.*" Others in the *Hatteras* thought the reply was: "H.M.S. *Spitfire.*" It was in fact: "Her Majesty's Steamer *Petrel.*"[5]

The *Alabama* was so obviously British-built that Blake's suspicions were almost set at rest and he said he would send a boat to *Vixen,* as he supposed her to be. While the boat was being lowered the *Alabama* opened fire. Although the *Hatteras* never had a chance, Blake fought her until she was on the verge of sinking in the hope of disabling the *Alabama* by a lucky shot and of attracting the attention of some of the other Union vessels 28 miles away. None of them came to his help and he and his crew were taken on board the *Alabama* barely 10 minutes before the *Hatteras* sank.

After fighting the *Hatteras* the *Alabama* cruised off

5. F. Moore, *Rebellion Record* (New York: G. P. Putnam; D. Van Nostrand, 1861–67), 6, doc. 357, 359; 10, doc. 226.

the coast of Brazil, around the Cape of Good Hope, back to Brazil, and put into Cherbourg, France, on June 11, 1864.

The United States consul at Cherbourg sent word by telegraph of the *Alabama*'s arrival to the minister at Paris; he relayed the news to Captain John A. Winslow, commanding the U.S.S. *Kearsarge,* which was at Antwerp, Belgium. Three days later the *Kearsarge* stood far enough into Cherbourg to make sure the *Alabama* was still there, then she took a station outside of the harbor.

Instead of trying to slip past the *Kearsarge* some dark night Semmes sent word through a Confederate agent and the United States consul to Winslow that if he would wait for her the *Alabama* would come out ready to fight. Winslow had been hunting for the *Alabama* for a long time; he was prepared to wait.

Semmes's decision to risk an almost irreplaceable Confederate vessel against one whose loss would be of no great importance to the Union was both foolish and vainglorious. He never explained why he acted as he did. However, any or all of several considerations could have motivated him. Probably he was imbued, as many southerners were, with exaggerated ideas of the sort of conduct demanded of a chivalrous man (which, of course, he thought himself to be). Perhaps he was tired of hearing the *Alabama* politely described as a mere commerce raider, impolitely as a pirate, and he felt it necessary to prove that she was a real warship. There is also reason to think he was told by some French naval officers (whose advice he believed would have been echoed by their British counterparts) that they would fight the *Kearsarge* if they had been in his place.

In any case, the *Alabama* steamed out of Cherbourg about 10 A.M., June 19, a fine Sunday with a light westerly breeze blowing. Partly to avoid any imputation of not being in waters beyond the jurisdiction of France, partly to prevent the *Alabama* from escaping to France if Semmes thought she might be defeated, the *Kearsarge* stood out to sea followed by the *Alabama* as though the latter were chasing the former. When the *Kearsarge* reached a point that suited Winslow's purpose, at 10:50 A.M., she turned on the *Alabama*. A few minutes later the *Alabama* opened fire with her starboard battery. The *Kearsarge* returned the compliment and tried to get into a raking position astern of her antagonist. The *Alabama* sheered and attempted to cross the bow of the *Kearsarge*. As the ships maneuvered to avoid being raked they revolved around a common center like partners in a deadly waltz; because there was a three knot westerly current they described a series of circles each a little to the west of the other.

By the time the seventh circle was completed nine of the *Alabama*'s men had been killed (including Dr. Llewellyn), 21 wounded, and the ship had been hulled a number of times. On receiving a report from his chief engineer that rising water had put out the fires Semmes ordered some sails set and tried to reach the French coast four and a half miles away, but it was already too late; the ship sank stern first a few minutes more than an hour after the battle began.

The *Kearsarge* saved 70 of the *Alabama*'s people and, at Winslow's request, the British yacht *Deerhound,* whose owner had come out to watch the fight, picked up 14 officers, including Semmes, and 26 seamen.

In March 1863 Commander Matthew Fontaine Maury bought the nearly new iron steamer *Japan* for the Confederate government.

Although the *Japan* was a speedy vessel, she was poorly suited for long cruises such as commerce raiders had to make because an iron hull quickly becomes foul unless it is protected with special paint (of a sort which did not exist in the 1860s). For this reason Bulloch strongly preferred wooden vessels whose hulls were covered with sheets of copper to keep them from being damaged by teredoes (shipworms). Maury was willing to take the *Japan* largely because wood was being superseded by iron in Great Britain and it was becoming difficult to buy or build wooden ships there.

The acquisition of the *Japan* was effected through the agency of a Liverpool company. One member of the firm posed as her owner and she was registered in the name of another member who shipped a crew of 50 men and cleared her for a voyage to Singapore. (At the urging of the American minister the two Britishers involved in these transactions were charged with having violated the Foreign Enlistment Act. They were tried and found guilty, but their punishment—fines of £50 each—can hardly be called severe.)[6]

The *Japan* sailed from Liverpool April 1 and rendezvoused a week later with the steamer *Alar* off Ushant, France. On April 9 she was armed and commissioned as C.S.S. *Georgia.*

After crossing the Atlantic a couple of times the *Georgia* put into Cherbourg. By this time her deficiencies had become so bothersome it was decided to shift her armament into another vessel. As things turned out this transfer was never made and she was sent to Liverpool to be sold. A local man bought her, reconverted her into a merchantman, and chartered her to the Portuguese government. En route to Lisbon she was captured by the U.S.S. *Niagara* and sent to the United States where she was condemned as a good prize. This action was upheld by the United States Supreme Court and her British owner lost the £15,000 he paid the Confederate government for her.[7]

---

6. Soley, *The Blockade,* pp. 214–15.

7. Ibid., p. 215; Bern Anderson, *By Sea and By River* (New York: Alfred A. Knopf, 1962), p. 210.

In November, 1863 Maury bought H.M.S. *Victor* to replace the unwanted *Georgia* and to be armed with her battery.

The *Victor,* a bark-rigged former Royal Navy dispatch boat, with a wooden hull, two steam engines, and a lifting propeller, seemed like an ideal cruiser; actually she had been condemned.

Maury, of course, sought to cover his tracks by having a British agent purchase the *Victor* and fit her out ostensibly for the China trade. But, as always happened, Adams heard of the deal and told the Foreign Office of his suspicions. By now the British no longer felt certain that the Confederates were going to win the war—the effectiveness of the Union blockade was beyond question, the Mississippi River was firmly controlled by Union forces, and the battle of Gettysburg (the high tide of the Confederacy) was a matter of recent history. In these circumstances orders were issued to detain the *Victor* until her real character could be determined. Nevertheless, she escaped from Sheerness, her Confederate officers joined her in the Thames River, and she was commissioned there as C.S.S. *Rappahanock.*

While the *Rappahanock* was still in the Thames estuary some of her bearings burned out and she was taken across the English Channel to Calais, France, for repairs. Moved by the same considerations as the British authorities had been, the French government found various pretexts for not letting the ship leave port. Finally, in March 1865, her captain officially turned her over to Bulloch, who tried to sell her. However, the days of the Confederacy were so clearly numbered that no one would buy her from him and she was eventually delivered to the United States government.

The cruiser *Tallahassee* began life as the *Atalanta* (or *Atlanta*), a twin-screw English Channel ferry. This ship had a reputation of being fast, yet comfortable, and is said to have "made the Dover-Calais crossing in 77 minutes on an even keel."[8]

Like many other British vessels the *Atalanta* became a blockade runner and she made at least two trips from Bermuda to Wilmington during the first six months of 1864.

In midsummer of that year she was bought by the Confederate Navy Department to be fitted out as a commerce raider. She departed from Wilmington under her new name on August 6 and made a cruise along the Atlantic coast. In 19 days she captured 33 coasters or fishing vessels and destroyed 26 of them; the others were bonded before being released. She could have done more if she had been as hospitably treated at Halifax, Nova Scotia, as most Confederate cruisers had been treated at other British ports. However, by the time she put into Halifax in need of coal the authorities there were paying more attention than they had in the past to the duties of a neutral power. She was supplied with just enough coal to reach Wilmington and she had to leave port after a stay of only 40 hours.

On her return to Wilmington the *Tallahassee* was renamed *Olustee* and sent on another cruise. This time she took and destroyed six vessels off the Delaware Capes before having to return to port for coal. The *Olustee*'s battery was now removed and she again became a blockade runner under the appropriate name of *Chameleon*. The *Chameleon* left Wilmington on December 24 while the Union fleet was preoccupied with the first attack on Fort Fisher. When she returned late in January 1865 she tried first to enter Wilmington, then Charleston.

8. *Civil War Naval Chronology* (Washington, D. C.: Naval History Division, Office of the Chief of Naval Operations, Navy Department, n. d.), pt. 6, p. 309.

Finding it impossible to make either of those ports she headed for Liverpool to be turned over to Bulloch. She reached there on April 9 and was seized by the British authorities who sold her. However, the United States government brought suit for her possession and she was delivered to the American consul at Liverpool on April 26.

In September 1864 Bulloch bought the full-rigged ship *Sea King* on her return from her maiden voyage to Bombay, India.

The *Sea King,* planked from keel to gunwale with teak fastened to iron frames, was rated A-1 for 14 years by Lloyds. Her logbook showed that on several occasions she had made more than 300 miles in 24 hours under sail alone and she had an 850-horsepower steam engine driving a lifting propeller.

Bulloch suggested to the Secretary of the Confederate Navy that the ship could advantageously be used to raid commerce, destroy troop transports, and lay "toll upon the exposed villages along the coast of New England," including "New Bedford, Holmes Hole [now Woods Hole], and Edgartown in Vineyard Sound, and Provincetown at the back of Cape Cod."[9] (On this occasion Bulloch displayed a southerner's ignorance of the geography of, and tactical situation in, Massachusetts where all of the places he named are located. New Bedford is not on Cape Cod, but on Buzzards Bay about 15 miles from Vineyard Sound and it was even then a fair-sized city. Moreover, with forts on both sides of the entrance to its harbor it would not have been helpless against an attack, particularly one by a single vessel.)

---

9. *Official Records of the Union and Confederate Navies in the War of the Rebellion* (Washington, D. C.: Government Printing Office, 1894–1922), Ser. 2, 2: 724.

Instead of following Bulloch's suggestion the Confederate Navy Department adopted a recommendation made by Commander Brooke to use the ship solely as a commerce raider with particular attention to the New England whaling fleet.

The *Sea King* cleared from London on October 8, ostensibly bound for Bombay, carrying coal as ballast (and to assure her of having enough fuel to stay at sea for a long time). On the same day the steamer *Laurel,* supposedly a blockade runner, sailed from Liverpool with 19 Confederate naval officers, posing as English passengers, and a cargo consisting of six cannon and supplies for a long voyage. These vessels met by arrangement near Funchal, Madeira, where, on October 19, the ship's new commanding officer, Lieutenant James I. Waddell of the Confederate States Navy, armed the *Sea King* and commissioned her as the C.S.S. *Shenandoah.*

Leaving Madeira, the *Shenandoah* cruised along the hitherto undisturbed route around the Cape of Good Hope to Australia, the South Pacific, the North Pacific, and the Arctic Oceans. On her way to the Cape she took six prizes, five of which were burned or scuttled and one was bonded and used to carry the crews of the others to Brazil.

After rounding the Cape the *Shenandoah* captured only one vessel before she put into Melbourne, Australia, on January 25, 1865. During the four weeks she stayed there she was fully refitted and allowed (with only token efforts to prevent it) to add a substantial number of men to her shorthanded crew.

On April 1, 42 days after leaving Melbourne, the *Shenandoah* found four whaleships at anchor in Lea Harbor at Ponape, or Ascension Island, in the Carolines. The captain of one of these vessels claimed she had been sold to Hawaiian interests, but Waddell did not believe him and his ship was burned with the others.

During the five weeks it took the *Shenandoah* to cross the rest of the Pacific Ocean and the Okotsk Sea she took only one prize, a New Bedford whaleship. However, she more than made up for lost time when she reached the Arctic Ocean late in June. Before she headed south again, on July 5, she took 21 more prizes, 11 of them being captured in seven hours.

Well satisfied with what he had accomplished so far Waddell returned to the Pacific planning to intercept shipping bound from the West Coast to Latin America or the Far East. On August 2 he learned from newspapers given to him by the captain of a British bark 13 days out of San Francisco that the Civil War had ended the previous April and he had literally been fighting for a lost cause. The *Shenandoah*'s guns were immediately dismounted and stowed in the hold, the hull was painted to make her look like an ordinary merchantman, and she set sail for Liverpool, 23,000 miles away, where Waddell hoped that he and his crew could find safety. She reached that port on November 6 and the British authorities turned her over to the United States government soon afterward.

The depredations of the Confederate cruisers led many American shipowners to sell their vessels to foreigners or to put them under British registry, with the result that the American merchant fleet dwindled rapidly. Because this was so it has often been said that the Confederate cruisers drove American shipping from the seas. Actually they did no more than to hasten an inevitable process. Twice before the Civil War—during the Revolutionary War and the War of 1812—the American merchant marine was "destroyed"; both times it quickly recovered because shipping offered attractive investment opportunities. With the opening of the West immediately after the Civil War ended, railroads, the manufacture of agricultural machinery, and meat packing, to mention only a few industries, offered more attractive investment opportunities than shipping did and money went into those fields rather than into the merchant marine.

16

Confederate Privateers

As little as the Confederate cruisers affected the outcome of the Civil War the Confederate privateers accomplished even less. However, their stories, to the extent they can be reconstructed from the scanty material available, are part of the record of the war.

The first vessel commissioned as a Confederate privateer was the 54-foot-long, 70-ton, screw tugboat *A. C. Gunnison*. The *Gunnison* was probably built at Philadelphia and is known to have been used in Troy, New York, as a Hudson River towboat. Armed with a spar torpedo containing 150 pounds of gunpowder and with her upper works protected by boiler plate, she was commissioned at Mobile on May 5, 1861. Apparently she served as a dispatch boat until November, 1863 when she was ordered, if possible, to attack the U.S.S. *Colorado* or one of the other vessels blockading Mobile. This mission was never accomplished and she was turned over to the Union Navy in April 1865.

The first privateer to make a cruise from a southern port and take any prizes was the 509-ton side-wheeler *Calhoun,* or *J. C. Calhoun,* as she was also called, of New Orleans.

Built in New York as the *Cuba,* this vessel was renamed, armed with an 18-pounder, two 12-pounders, and two 6-pounders, and commissioned May 15, 1862.

During a five-month cruise she took and sent into New Orleans the brig *Panama,* the schooners *Mermaid, John Adams,* and *Ella,* the bark *Ocean Eagle,* and the ship *Milan.*

On her return to New Orleans the *Calhoun* was chartered by the Confederate Navy. While thus employed she was captured by the U.S.S. *Colorado*'s tender, the schooner *Samuel Rotan,* off the Southwest Pass of the Mississippi River on January 23, 1862.

The worst possible luck plagued the third Confederate privateer, the schooner *Savannah* of Charleston.

This 56-foot-long, 53-ton vessel was originally Charleston harbor Pilot Boat No. 7. Armed with an 18-pound cannon cast in 1812, she sailed from Charleston June 3, 1861. Early the following morning she captured the brig *Joseph,* loaded with sugar. After escorting her prize into Georgetown, South Carolina, the *Savannah* put out to sea again. When another brig was sighted an hour or so later the crew, flushed with success, made all sail, expecting soon to bag another prize. However, the chase turned out to be the U.S.S. *Perry,* which, with her guns housed looked like a merchantman. By the time the *Savannah*'s people realized their mistake it was too late to escape and she was captured after a short fight.

With a view to discouraging other Confederate privateers the Union authorities sent the *Savannah*'s 32 officers and men to New York to be tried on charges of piracy, a capital crime. Their trial began on October 23, 1861, and lasted for eight days. Their counsel did not deny the facts in the case, but the judge was persuaded to charge the jury that if there had been no felonious intent (i.e. if there had been no design to take property merely for the sake of personal gain) the defendants could not be found guilty of piracy, no matter what other offense they might have committed. After deliberating for 24 hours the jury reported itself hopelessly deadlocked and was discharged.

The *Savannah*'s people were not brought to trial again because when they were first taken to court the Confederate government chose by lot 32 high-ranking Union officers who had been captured at the battle of Bull Run, or Manassas, and threatened to hang one of them for every privateersman executed. In these circumstances President Lincoln decided at the turn of the year that captured privateersmen were to be treated as ordinary prisoners of war. Nevertheless, northern shipping interests and most northern newspapers referred to all Confederate sailors, whether they were privateersmen or officers and seamen in duly commissioned naval vessels, as pirates throughout the war and long afterward.

The *Jefferson Davis,* a 187 foot long brig, was one of the comparatively successful Confederate privateers.

This vessel was built in Baltimore about 1845 as the *Putnam.* She engaged in the African slave trade under the name *Echo* until she was captured off Cuba with 271 Negroes on board by the U.S.S. *Dolphin,* commanded by Lieutenant Maffitt (who was afterward captain of the C.S.S. *Florida*). The *Echo* was condemned and sold to Captain Robert Hunter who restored her original name. When the war began he formed a syndicate which included some of the elite of Charleston and applied for a privateering commission.

Armed with two 32-pounders, two 24-pounders, and an 18-pounder, all made in England in 1801, the *Jefferson Davis* sailed from Charleston on June 28, 1862, with a crew of 75 officers and men.

She took her first prizes, the brig *John Welsh,* bound

for Falmouth, England, with a cargo of sugar, and the schooner *Enchantress,* from Boston to Santiago, Cuba, off Cape Hatteras on July 6.

A day later, 150 miles from Montauk Point at the end of New York's Long Island, the *Jefferson Davis* captured the schooner *J. D. Waring,* bound from Brookhaven, New York, for Montevideo, Uruguay, with a general cargo. The schooner was ordered into Charleston in charge of a prize crew, but she never reached there. While she was about 50 miles from that port her cook killed the prize master and his two mates with an ax. The few other Confederates surrendered and helped work the vessel to Long Island.

On July 9 the *Jefferson Davis* took the brig *Mary E. Thompson,* from Searsport, Maine, for Montevideo, with a cargo of lumber, and the ship *Mary Goodell,* from New York for Buenos Aires, Argentina. The brig was sent to Charleston, but the ship's draught was too great to permit her to enter that harbor so she was released on bond and sailed for Portland, Maine, carrying the *Jefferson Davis*'s prisoners with her.

Heading toward South America, the *Jefferson Davis* took the bark *Alvarado,* bound from the Cape of Good Hope for Boston with a cargo of hides, wool, etc., on July 21. After capturing the schooner *Windward,* carrying salt she had loaded at Turk's Island, the *Jefferson Davis* headed north again and fell in with the ship *John Crawford* of Philadelphia with arms and coal consigned to the Union force at Key West, Florida. Because the ship drew 22 feet her people were transferred to the privateer and she was set on fire.

A short time later, on August 17, the *Jefferson Davis* was wrecked as she tried to cross the bar at the mouth of the harbor of St. Augustine, Florida.

The briefest career of any Confederate privateer was

that of the schooner *Petrel.* Originally the Charleston harbor Pilot Boat *Eclipse,* this vessel was purchased by the Revenue Cutter Service (the forerunner of the Coast Guard) and became the Cutter *William Aiken* which was seized by the state of South Carolina at the beginning of the Civil War.

Commissioned at Charleston on July 10, 1861, the *Petrel* sailed from that port on the 28th of the same month. Before the day ended she was overhauled by the U.S.S. *St. Lawrence* after a four-hour chase. The *Petrel* managed to fire only three shots during the 30 minutes it took the *St. Lawrence* to sink her with two hits. Two men went down with the *Petrel;* the rest of her crew was saved by the *St. Lawrence.*

A 518 ton side-wheeler with a speed of 16 knots first named *Caroline,* then *Gordon, Theodora,* and finally *Nassau,* mixed blockade running, privateering, and service under charter by the Confederate States Navy. (As already mentioned she was chartered to take Mason and Slidell to Cuba on the first leg of their journeys to Great Britain and France.)

The *Caroline* was built at Greenpoint, on the East River in Brooklyn, New York, for the Florida Steamship Company, which ran her between Charleston and Fernandina with occasional crossings to Cuba.

Armed with three guns, this vessel was commissioned as the privateer *Gordon* at Charleston on July 15, 1861. Under that name she took the brig *William McGilvery,* bound from Cardenas to Bangor, Maine, with a cargo of molasses, on July 25 and the schooner *Protector,* bound from Cuba for Philadelphia with fruit, three days later.

The *Gordon* soon turned from privateering to the more lucrative business of blockade running and on May 28, 1863, now known as the *Nassau,* she was destroyed by the U.S.S. *Victoria* while trying to enter the Cape Fear River with a cargo of Enfield rifles, ammunition, clothing, and medicines for the Confederate Army.

Another short lived privateer was the brig *Beauregard,* built as the schooner *Priscilla C. Ferguson,* for Charleston owners. Commissioned on November 14, 1861, the *Beauregard* sailed from Charleston the following day. A week later she was surprised by the U.S.S. *W. G. Anderson,* whose attention was attracted by the large number of men on the *Beauregard*'s deck. When she was overhauled after a two-hour chase her crew threw all of the arms and ammunition overboard and cut up her sails and rigging before surrendering. Brought into Key West on November 19, the *Beauregard* was taken into the Union Navy and, with her name unchanged, she served with the Eastern Gulf Squadron during the rest of the war.

The last Confederate privateer of which any record has been found was the 110-foot-long schooner *Dixie.*

This craft, built in Baltimore as the *H. and J. Neild,* was engaged in the West Indies trade at the beginning of the Civil War. She was then renamed *Dixie* and became a blockade runner. Commissioned as a privateer at Charleston on June 21, 1863, the *Dixie* sailed from that port on July 20. She took her first prize, the bark *Glenn* of Portland, Maine, loaded with coal, on July 23. Her next capture, the schooner *Mary Alice,* was made on July 25. The *Mary Alice* was ordered into Morehead City, but was retaken by the U.S.S. *Wabash.* A few days later the *Dixie* chased two vessels which escaped in a squall. Then she spoke the *Robert B. Kirkland.* Finding that she was from Baltimore, the *Dixie* let her go with the *Dixie*'s prisoners on board. The bark *Rowena* of Philadelphia, bound home with a cargo of sugar, was taken by the *Dixie* on July 31. The *Rowena* had a large crew so the *Dixie*'s captain took charge of her, leaving only five men in the schooner. After some narrow escapes the *Dixie* and her prize made their way into Charleston via Bull's Bay. A few months later the *Dixie* was sold to a Charleston firm, which renamed her *Kate Hale* and then *Success*.

17
How the North Won the War

It is no exaggeration to say that if the Union Navy had not possessed 500 or 600 vessels to the Confederate Navy's less than 150 the North would not have won the Civil War.

The most important use the Union made of its overwhelming naval superiority, and one to which historians have paid too little attention, was the establishment and maintenance of the blockade of the Confederate States.

During the war the Union Navy captured 1149 blockade runners, 210 of them steamers, and burned, sank, or drove ashore another 335 vessels, including 85 steamers. These craft and their cargoes were worth at least $31 million at a time when the purchasing power of the dollar was vastly greater than it is now. There was also the substantial, though inestimable, effect of the discouragement to trade—of cargoes not shipped—because of the blockade. To a country as dependent on the outside world for as many things as the Confederacy was these were fatal blows.

Because the blockade was not airtight some authors, notably Frank L. Owsley in his *King Cotton Diplomacy* and Francis B. C. Bradlee in *Blockade Running During the Civil War,* have suggested that it had no effect on the outcome of the war. No southerner who lived through the war would have agreed with this thesis. They knew that their side was winning until 1863, when the blockade made the Confederate States into a land besieged, a land unable any longer to procure the matériel with which to continue the war or even to obtain the very necessities of life. They knew, as a contemporary southerner, J. Thomas Scharf, said in the preface to his *History of the Confederate Navy,* that the blockade "shut the Confederacy out from the world, deprived it of supplies, weakened its military and naval strength, . . . compelled exhaustion by requiring the consumption of everything grown or raised in the country," and finally determined the outcome of the war.

The Union Navy won singlehanded victories only at Port Royal, New Orleans, and Memphis, but few historians have appreciated the extent to which it helped the Army. Its ability to guarantee the safe transportation of troops anywhere on the southern coasts or rivers greatly weakened the Confederate armies in the field by making it necessary for men to be stationed at many places that actually were never attacked. Significantly, too, the Union won most of the battles fought where the two services could cooperate (e.g. everywhere on the western rivers and Roanoke Island in the East) while those fought where naval assistance was impossible (e.g. First and Second Bull Run, Fredericksburg) often resulted in Union defeats or stalemates, as at Antietam and Gettysburg.

Bibliography

This book is based primarily on the following sources:

Ammen, Daniel. *The Atlantic Coast.* New York: Charles Scribner's Sons, 1883.

Mahan, A. T. *The Gulf and Inland Waters.* New York: Charles Scribner's Sons, 1883.

Naval History Division, Office of the Chief of Naval Operations, Navy Department. *Civil War Naval Chronology* (six parts). Washington, D. C., no date.

Scharf, J. Thomas. *History of the Confederate States Navy.* New York: Rogers & Sherwood, 1887.

Soley, James Russell. *The Blockade and the Cruisers.* New York: Charles Scribner's Sons, 1883.

U. S. Office of Naval Records. *Official Records of the Union and Confederate Navies in the War of the Rebellion* (30 vols.). Washington, D. C.: Government Printing Office, 1894–1912.

Other works consulted, and in some cases cited, include

Abbot, Willis J. *Blue Jackets of '61.* New York: Dodd, Mead, and Company, 1887.

Abbott, John S. C. "Charles Ellet and His Naval Steam Rams." *Harper's New Monthly Magazine* 32 (February 1866): pp. 295–312.

"The Navy in the North Carolina Sounds." *Harper's New Monthly Magazine* 32 (April 1866):

"Opening the Mississippi." *Harper's New Monthly Magazine* 33 (August 1866):

Adams, Charles Francis. *Charles Francis Adams.* Boston: Houghton, Mifflin and Company, 1900.

Adams, Ephraim D. *Great Britain and the American Civil War* (two volumes bound as one). New York: Russell & Russell, Inc., 1958.

Ambrose, Stephen E.; Bearss, Edwin C.; Nye, Wilbur S.; Ray, Frederic; and Utley, Beverly. "Struggle for Vicksburg." *Civil War Times,* July 1967, entire issue.

Ammen, Daniel, *The Old Navy and the New.* Philadelphia: J. B. Lippincott, 1891.

Anderson, Bern. *By River and By Sea.* New York: Alfred A. Knopf, 1962.

Baxter, James Phinney, 3rd. *The Introduction of the Ironclad Warship.* Cambridge: Harvard University Press, 1933.

Bearss, Edwin C. *Hardluck Ironclad.* Baton Rouge, La.: Louisiana State University Press, 1966.

Bearss, Edwin C., and Nash, Howard P., Jr. "The Attack on Fort Henry." *Civil War Times,* November 1965, pp. 9–15.

Bennett, Frank M. *The Steam Navy of the United States.* Pittsburgh: Warren & Co., 1896.

———. *The Monitor and the Navy Under Steam.* Boston: Houghton, Mifflin and Company, 1900.

Blair, Lieutenant Carval H. "Submarines in the Confederate Navy." *United States Naval Institute Proceedings,* October 1952.

Bolitho, Hector. *Albert Prince Consort.* Indianapolis: The Bobbs-Merrill Company, Inc., 1964.

Boykin, Edward. *Ghost Ship of the Confederacy.* New York: Funk & Wagnalls, 1957.

Boynton, Charles B. *History of the Navy During the Rebellion* (two vols.). New York: D. Appleton and Company, 1867, 1868.

Bradlee, Francis B. C. *Blockade Running During the Civil War.* Salem, Mass.: The Essex Institute, 1925.

Brooke, George Mercer, Jr. *John Mercer Brooke.* Unpublished thesis, University of North Carolina, Chapel Hill, N. C., 1955.

Bruce, Robert V. *Lincoln and the Tools of War*. Indianapolis: The Bobbs-Merrill Company, Inc., 1955.

Bulloch, James D. *The Secret Service of the Confederate States in Europe* (two vols.). London: Richard Bentley and Sons, 1883.

Carese, Robert. *Blockade*. New York: Rinehart &. Co., Inc., 1958.

Carrison, Daniel J. *The Navy from Wood to Steel*. New York: Franklin Watts, Inc., 1965.

Church, William Conant. *The Life of John Ericsson* (two volumes bound as one). New York: Charles Scribner's Sons, 1911.

Cochran, Hamilton. *Blockade Runners of the Confederacy*. Indianapolis: The Bobbs-Merrill Company, Inc., 1958.

Crafts, W. A. *The Southern Rebellion* (two vols.). Boston: Samuel Walker & Co., 1866, 1867.

Crandall, Warren D., and Newell, Isaac D. *History of the Ram Fleet*. St. Louis: Buschart Brothers, 1907.

Cranwell, John Phillips. *Spoilers of the Sea*. New York: W. W. Norton & Company, 1941.

Dahlgren, Madeline V. *Memoir of John A. Dahlgren*. Boston: James R. Osgood and Company, 1882.

Daly, R. W. *How the Merrimac Won*. New York: Thomas Y. Crowell Company, 1957.

Daly, Robert W. (editor). *Aboard the USS* Monitor: *1862*. Annapolis, Md.: United States Naval Institute, 1964.

Dalzell, George W. *The Flight from the Flag*. Chapel Hill, N. C.: University of North Carolina Press, 1940.

Davis, Charles H. *Life of Charles Henry Davis*. Boston: Houghton, Mifflin and Company, 1899.

Dowdey, Clifford. *Experiment in Rebellion*. New York: Doubleday & Company, Inc., 1946.

Eisenschmil, Otto, and Newman Ralph. *The Civil War*. New York: Grossett & Dunlap, Inc., 1956.

Farragut, Loyall. *The Life of David Glasgow Farragut.* New York: D. Appleton and Company, 1879.

Gosnell, H. P. *Guns on Western Waters.* Baton Rouge, La.: Louisiana State University Press, 1949.

Grant, U. S. *Personal Memoirs of U. S. Grant* (two vols.). New York: Charles L. Webster & Company, 1885–86.

Greene, F. V. *The Mississippi.* New York: Charles Scribner's Sons, 1883.

Headley, J. T. *The Great Rebellion* (two vols.). Hartford, Conn.: American Publishing Company, 1866.

———. *Farragut and Our Naval Commanders.* New York: E. B. Treat & Co., 1867.

Hill, Jim Dan. *Sea Dogs of the Sixties.* Minneapolis, Minn.: University of Minnesota Press, 1935.

Holden, Edgar. "The First Cruise of the Monitor *Passaic.*" *Harper's New Monthly Magazine,* October 1863, pp. 578–94.

Hoppin, James Mason. *Life of Andrew Hull Foote.* New York: Harper & Brothers, 1874.

Horner, David. *The Blockade Runners.* New York: Dodd, Mead and Company, 1968.

Huffstot, Robert D. "The *Carondolet* and Other 'Pook Turtles.'" *Civil War Times,* August 1967, pp. 5–11, 46–48.

———. "Battle for the Post of Arkansas." *Civil War Times,* January 1969, pp. 10–19.

Hunt, Cornelius N. *The Shenandoah.* New York: C. W. Carleton & Co., 1867.

Johns, John. "Wilmington During the Blockade." *Harper's New Monthly Magazine,* September 1866, pp. 497–502.

Johnson, Robert U., and Buel, Clarence (editors). *Battles and Leaders of the Civil War* (four vols.). New York: The Century Co., 1884–87.

Jones, Virgil C. *Civil War at Sea* (three vols.). New York: Holt, Rinehart, Winston, 1961–62.

Jones, V. C., "How the South Created a Navy." *Civil War Times,* July 1969, pp 4–9, 42–48.

Lewis, Charles Lee. *Admiral Franklin Buchanan.* Baltimore,

Md.: The Norman, Remington Company, 1929.

———. *David Glasgow Farragut* (Vol. II). Annapolis, Md.: United States Naval Institute, 1943.

Lewis, Gene D. *Charles Ellet, Jr.* Urbana, Ill.: University of Illinois Press, 1968.

Lewis, Paul. *Yankee Admiral.* New York: David McKay Company, Inc., 1968.

Lockwood, H. C. "The Capture of Fort Fisher." *The Atlantic Monthly,* May 1871, pp. 622–36.

Lossing, Benson J. *The Civil War in America* (three vols.). Philadelphia: George W. Childs, 1866, 1868, 1870.

Luvaas, Jay. "Burnside's Roanoke Island Campaign." *Civil War Times,* December 1968, pp. 4–11, 43–48.

Macartney, Clarence Edward. *Mr. Lincoln's Admirals.* New York: Funk & Wagnalls, 1956.

MacBride, Robert. *Civil War Ironclads.* Philadelphia: Chilton Books, 1962.

Maclay, Edgar Stanton. *A History of the United States Navy* (Vol. II.). New York: D. Appleton and Company, 1894.

———. *Reminiscences of the Old Navy.* New York: G. P. Putnam's Sons, 1898.

Mahan, A. T. *Admiral Farragut.* New York: D. Appleton and Company, 1892.

Mahon, John K. "The Civil War Letters of Lieutenant Commander George Bacon." *American Neptune,* October 1952, pp. 271–81.

Marshall, Jesse Ames (compiler). *Private and Official Correspondence of Gen. Benjamin F. Butler During the Period of the Civil War.* Privately Issued, 1917.

Martin, Christopher. *Damn the Torpedoes.* New York: Abelard-Schuman, 1970.

Matthews, Franklin. *Our Navy in Time of War.* New York: D. Appleton and Company, 1909.

Melton, Maurice. *The Confederate Ironclads.* South Brunswick, N. J.: Thomas Yoseloff, 1968.

Merrill, James M. *The Rebel Shore*. Boston: Little, Brown and Company, 1957.

Miller, F. T., and Barnes, J. (editors). *Photographic History of the Civil War* (Vol. VI). New York: Review of Reviews, 1911.

Millis, Walter. "The Iron Sea Elephants." *American Neptune*, January 1950, pp. 15–32.

Moore, F. *Rebellion Record* (12 vols.). New York: G. P. Putnam; D. Van Nostrand, 1861–67.

Morgan, Murray. *Dixie Raider*. New York: E. P. Dutton & Co., Inc., 1948.

Nash, Howard P., Jr. "Ironclads at Charleston." *Civil War Times*, January 1960, pp. 1–4.

———. "Ironclads, Tinclads and Cottonclads." *Tradition*, October 1960, pp. 6–16.

———. "A Civil War Legend Examined." *American Neptune*, July 1963, pp. 197–203.

———. "The CSS *Alabama:* Roving Terror of the Seas." *Civil War Times*, August 1963, pp. 6–9, 34–39.

———. "The Ignominious Stone Fleet." *Civil War Times*, June 1964, pp. 43–49.

———. "Shenandoah." New Bedford, Massachusetts. *Sunday-Standard Times*, June 20, 1965.

———. "The Story of Island No. 10." *Civil War Times*, December 1966, pp. 42–47.

———. *Stormy Petrel*. Rutherford, N. J.: Fairleigh Dickinson University Press, 1969.

Nicolay, John G. *The Outbreak of the War*. New York: Charles Scribner's Sons, 1883.

Nicolay, John G., and Hay, John. *Abraham Lincoln* (Vols. IV–X). New York: The Century Co., 1890.

Osbon, B. S. *Hand Book of the United States Navy*. New York: D. Van Nostrand, 1864.

Parker, John C. "With Farrgut at Port Hudson." *Civil War Times*, November 1968, pp. 42–49.

Peck, W. G. F. "Four Years Under Fire at Charleston." *Harper's New Monthly Magazine,* August 1865, pp. 358–66.

Porter, David D. *Incidents and Anecdotes of the Civil War.* New York: D. Appleton and Company, 1885.

———. *Naval History of the Civil War.* New York: The Sherman Publishing Company, 1886.

Pratt, Fletcher. *Ordeal by Fire.* New York: William Sloane Associates, 1935.

———. *Civil War on Western Waters.* New York: Henry Holt and Company, 1956.

———. *Compact History of the United States Navy.* New York: Hawthorn Books, Inc., 1957.

Price, Marcus W. "Ships That Tested the Blockade of the Carolina Ports, 1861–1865." *American Neptune,* July 1948, pp. 00-00.

———. "Blockade Running as a Business in South Carolina." *American Neptune,* January 1949, pp. 31–62.

Rae, Thomas. "The Little Monitor Saved Our Lives." *American History Illustrated,* July 1966, pp. 32–39.

Rhodes, James Ford. *History of the Civil War.* New York: The Macmillan Company, 1917.

Robinson, William Morrison, Jr. *The Confederate Privateers.* New Haven, Conn.: Yale University Press, 1928.

Roske, Ralph J., and Van Doren, Charles. *Lincoln's Commando.* New York: Harper & Brothers, 1957.

Semmes, Raphael. *Memoirs of Service Afloat.* Baltimore, Md.: Kelly, Piet & Co., 1869.

Sinclair, Arthur. *Two Years on the Alabama.* Boston: Lee & Shepard, 1896.

Soley, James Russell. *The Sailor Boys of '61.* Boston: Estes and Lauriat, 1888.

———. *Admiral Porter.* New York: D. Appleton and Company, 1903.

Spears, John Randolph. *David G. Farragut.* Philadelphia: George W. Jacobs & Company, 1905.

Sprout, Harold and Margaret. *The Rise of American Naval*

Power. Princeton, N. J.: Princeton University Press, 1967.

Stephenson, Nathaniel W. *Abraham Lincoln and the Union*. New Haven, Conn.: Yale University Press, 1920.

———. *The Day of the Confederacy*. New Haven, Conn.: Yale University Press, 1920.

Thompson, Robert Means, and Wainwright, Richard (editors). *Confidential Correspondence of Gustavus Vasa Fox* (two vols.). New York: Printed for the Naval History Society by the De Vinne Press, 1918–19.

Vickers, George Marley (editor). *Under Both Flags*. Grand Rapids, Mich.: P. D. Farrell Company, 1896.

Welles, Gideon. *Diary of Gideon Welles* (three vols.). Boston: Houghton, Mifflin and Company, 1911.

———. *Letter of the Secretary of the Navy . . . in Relation to the construction of the ironclad* Monitor. Washington, D. C.: Government Printing Office, 1868.

West, Richard S., Jr. *The Second Admiral*. New York: Coward-McCann, Inc., 1937.

———. *Gideon Welles*. Indianapolis: The Bobbs-Merrill Company, 1943.

———. *Mr. Lincoln's Navy*. New York: Longmans Green and Company, 1957.

White, William C. and Ruth. *Tin Can on a Shingle*. New York: E. P. Dutton & Company, 1957.

Williams, George F. *The Memorial War Book*. New York: Dreyfus Publishing Co., 1894.

Wilson, H. W. *Ironclads in Action* (two vols.). Boston: Little, Brown and Company; London: Sampson Low, Marston and Company, 1896.

Wilson, John Laird. *The Pictorial History of the Great Civil War*. Philadelphia: The National Publishing Co., 1878.

Wood, William. *Captains of the Civil War*. New Haven, Conn.: Yale University Press, 1921.

Woodburn, James A. *The Trent Affair*. Indianapolis and Kansas City: The Bobbs-Merrill Company, 1896.

Woodward, W. E. *Meet General Grant*. New York: Horace Liveright, Inc., 1928.

Index